Living with the Texas shore

Living with the shore

Series editors
Orrin H. Pilkey, Jr.
William J. Neal

Editorial consultant
Martin Wilcox

Living with the Texas shore

Robert A. Morton
Orrin H. Pilkey, Jr.
Orrin H. Pilkey, Sr.
William J. Neal

Duke University Press   Durham, North Carolina   1983

The publication of this volume was subsidized in part by gifts from M. Harvey Weil and Patrick H. Welder, Jr.

© 1983 Duke University Press, all rights reserved

Printed in the United States of America

Library of Congress Cataloging in Publication Data
Main entry under title:

Living with the Texas shore

   (Living with the shore)
   Bibliography: p.
   Includes index.
   1. Seashore—Texas—Gulf Region. 2. Barrier islands
—Texas—Gulf Region. I. Morton, Robert A. II. Series.
GB459.25.L58   1983   333.91'7'09764   83-1753
ISBN 0-8223-0499-6
ISBN 0-8223-0500-3 (pbk.)

Contents

Foreword xi

1. A coastal perspective 3

A brief description of the Texas coast 3
Key events in the history of the coast 6
    The first settlers 6
    Early ports and forts 6
    Resort communities 7
    High-density development 7
    Hurricane history: a stormy past 9

2. Dynamics of the Texas shore 12

Barrier islands: how they were formed 12
The operation of barrier islands 16
    How islands differ 16
    How islands are similar 17
    Barrier island environments are interrelated 19
Beaches: the dynamic equilibrium 20
    How does the beach respond to a storm? 20
    Where does the beach sand come from? 21
    How does the beach widen? 23
    Where do seashells come from? 23
    Why do beaches erode? 23
    How can I tell if my shoreline is eroding? 24
    If most of the Gulf shorelines are eroding, what is the long-range future of Texas development? 24
    Are the shorelines on the back sides of Texas islands eroding? 24
    What can I do about my eroding beach? 24
Hurricane origin: a storm is born 25
Hurricane forces: wind, waves, and washover 26

3. Man and the shoreline 29

Stabilizing the unstable 29
    Beach replenishment 29
    Groins and jetties 30
    Seawalls 31
A philosophy of shoreline conservation: "We have met the enemy and he is us" 34
Truths of the shoreline 35
The solutions 38

4. Selecting a site on a Texas beach 39

Nature's clues to danger at the beach 40
    Elevation 40
    Vegetation 41
    Dunes 41
    Soils 41
    Terrain 42

The problems people cause at the beach 45
    Stabilization of shorelines 45
    Construction 45
    Water and sewage 46
    Finger canals 46
    Escape routes 48
The site: checklist for safety evaluation 49
Site analysis: the coast of Texas 50
    Sabine Pass to the vicinity of Sea Rim State Park 50
    Sea Rim State Park vicinity to High Island vicinity 53
    High Island vicinity to Rollover Pass 53
    Rollover Pass to Crystal Beach vicinity 56
    Crystal Beach vicinity to Bolivar Roads 57
    Galveston Island 62
    Galveston jetty to Galveston seawall 63
    Galveston seawall 69
    Galveston seawall to vicinity of Eleven-Mile Road 69
    Eleven-Mile Road to San Luis Pass 70
    San Luis Pass to Surfside 71
    Surfside to the Brazos River 71
    Brazos River to the San Bernard River 76
    San Bernard River to Cedar Lakes 76
    Cedar Lakes to Sargent Beach 77
    Sargent Beach to Brown Cedar Cut 77
    East Matagorda Peninsula to Spring Bayou 77
    Spring Bayou to the Colorado River 80
    West Matagorda Peninsula to Greens Bayou 80

    Greens Bayou to Matagorda Ship Channel 86
    Matagorda Ship Channel to Pass Cavallo 86
    Matagorda and San Jose Islands 87
    East end of Matagorda Island 87
    Air Force Base to Power Lake vicinity 87
    Power Lake vicinity to Panther Point vicinity 92
    Panther Point vicinity to Cedar Bayou 92
    Cedar Bayou to North Pass 92
    North Pass to Aransas Pass 97
    Aransas Pass to Mustang Island State Park 97
    Corpus Christi Pass to Packery Channel 101
    Packery Channel to Padre Island National Seashore 104
    Mansfield Channel to Veterans Park 105
    Veterans Park to Brazos Santiago Pass 105
    Brazos Santiago Pass to State Highway No. 4 109
    State Highway No. 4 to mouth of the Rio Grande 109

5. The barrier coast of Texas, land use, and the law 115

The National Flood Insurance Program 115
Coastal Barriers Resources Act 116
Waste disposal 117
Building codes 117
Open beaches 118
Excavation of sand 118
Sand dune protection 118
The Texas Coastal Public Lands Management Act 119
The Texas Coastal Program 120

6. Building or buying a house near the beach   121
   Coastal realty versus coastal reality   121
   The structure: concept of balanced risk   122
   Coastal forces: design requirements   123
      Hurricane winds   123
      Storm surge   125
      Hurricane waves   125
      Barometric pressure change   125
   House selection   126
   Keeping dry: pole or "stilt" houses   126
   An existing house: what to look for, where to improve   131
      Geographic location   131
      How well built is the house?   132
      What can be done to improve an existing house?   137
   Mobile homes: limiting their mobility   139
   High-rise buildings: the urban shore   142
   Modular unit construction: prefabricating the urban shore   145
   An unending game: only the players change   147

Appendix A. Hurricane checklist   148

Appendix B. A guide to federal, state, and local agencies involved in costal development   152

Appendix C. Useful references   160

Appendix D. Field trip guides   175

   Index   185

Figures and tables

Figures

1.1. Index map of Texas barrier islands and beaches   4
1.2. High-density development, South Padre Island   8
1.3. Undeveloped barrier segment, Mustang Island   8
1.4. Overdevelopment of the New Jersey coast   9
1.5. Historic shoreline changes at Sargent Beach   9
1.6. Hurricane tracks of the sixties and seventies   10

2.1. 17,000 years of sea-level change   13
2.2. Evolution of a barrier island   14
2.3. Ratio of horizontal island migration to vertical sea-level rise   15
2.4. Barrier island environments   19
2.5. The dynamic equilibrium of the beach   20
2.6. Beach flattening in response to a storm   22
2.7. Washover channels cut through Matagorda Peninsula by Hurricane Carla in 1961   27

3.1. Beach nourishment   30
3.2. Groined shoreline   32
3.3. Saga of a seawall   33
3.4. Cape May, New Jersey, seawall (1976)   35
3.5. Galveston seawall   36
3.6. Failed seawall on South Padre Island   37
3.7. Monmouth Beach, New Jersey, seawall   38

4.1. Dune migration on North Padre Island   41
4.2. Overwash from Hurricane Carla (1961) on Bolivar Peninsula   42
4.3. Overwash of road by Hurricane Beulah in 1967   43
4.4. Finger-canal problems   47
4.5. Queen Isabella Causeway, South Padre Island   49
4.6. Site analysis: Gulf beach near Sabine Pass   51
4.7. Site analysis: Gulf beach near Sea Rim State Park   52
4.8. Site analysis: Sea Rim State Park to High Island   54
4.9. Road destroyed by beach erosion near High Island   56
4.10. Site analysis: Bolivar Peninsula from Gilchrist to Caplen   58
4.11. Site analysis: Bolivar Peninsula from Crystal Beach to Bolivar Roads   60
4.12. Wellhead in surf zone of Bolivar Peninsula   62
4.13. Site analysis: Galveston Island from Bolivar Roads to end of the seawall   64
4.14. Site analysis: Galveston Island from end of seawall to Sea Isle   66
4.15. Site analysis: western end of Galveston Island and eastern end of Follets Island   68
4.16. Retreating beach, west end of Galveston seawall   69
4.17. Sand borrow pit, West Beach, Galveston Island   70
4.18. Site analysis: eastern end of Follets Island to Quintana   72
4.19. Site analysis: Bryan Beach to Cedar Lakes   74
4.20. Beach house on an eroding beach, Surfside   76

4.21. Site analysis: Cedar Lakes to Brown Cedar Cut   78
4.22. Outcrop of marsh mud, Sargent Beach   80
4.23. Site analysis: East Matagorda Peninsula in the vicinity of Brown Cedar Cut   81
4.24. Site analysis: East and West Matagorda Peninsula in the vicinity of the Colorado River   82
4.25. Site analysis: West Matagorda Peninsula to Pass Cavallo   84
4.26. Site analysis: Matagorda Island in the vicinity of Pass Cavallo   88
4.27. Site analysis: Matagorda Island from Power Lake to Cedar Bayou   90
4.28. Site analysis: San Jose Island in the vicinity of Cedar Bayou   93
4.29. Site analysis: San Jose Island from Mud Island to Aransas Pass   94
4.30. Migration of Aransas Pass   96
4.31. Site analysis: Mustang Island from Aransas Pass to Mustang Island State Park   98
4.32. Condominiums on Mustang Island   100
4.33. Port Aransas trailer park after Hurricane Celia (1970)   101
4.34. Site analysis: Mustang Island and North Padre Island from Corpus Christi Pass to Padre Island National Seashore   102
4.35. Failed seawall on North Padre Island   104
4.36. Site analysis: South Padre Island from Mansfield Channel to the Willacy-Cameron county line   106
4.37. Site analysis: South Padre Island near Willacy-Cameron county line   108
4.38. Site analysis: South Padre Island from La Punta Larga to Veterans Park   110
4.39. Site analysis: South Padre Island and Boca Chica Island to the Rio Grande   112
4.40. Condominium on South Padre Island and similar structure at Panama City damaged by Hurricane Eloise in 1975   114

6.1. Forces to be reckoned with at the seashore   124
6.2. Modes of failure and how to deal with them   127
6.3. Shallow and deep supports for poles and posts   128
6.4. Pole house with poles extending to roof   129
6.5. Framing system for an elevated house   130
6.6. Tying floors to poles   131
6.7. Foundation anchorage   133
6.8. Stud-to-floor, plate-to-floor framing methods   133
6.9. Roof-to-wall connections   134
6.10. Where to strengthen a house   135
6.11. Reinforced tie beam (bond beam) for concrete block walls   137
6.12. Tiedowns for mobile homes   140
6.13. Hardware for mobile-home tiedowns   141
6.14. Some rules in selecting or designing a house   146

Tables

1.1. Probability of storms (the number expected over a 100-year period)   11
2.1. Still-water surge levels for twentieth-century hurricanes   26
2.2. Summary of natural hazards and their impact on the Texas coast   27
6.1. Tiedown anchorage requirements   142

Foreword

"The world is too much with us" and we all seek relief. In Texas, the beaches are more and more *the* place for relief and Texans are rushing to the shore in record-breaking numbers. No wonder! The Texas shore is one of the greatest recreational assets of a state blessed with many recreational resources. It is also a fact that the Texas shoreline is the most varied in America.

But there is trouble in paradise. Texans love their shore too much. Not satisfied with a few simple days of frolicking in the sun and surf, many Texans are leaving behind cottages and condominiums, second homes built as close to the beach as possible.

Texans, or at least those with a sense of history, know better than any other Americans the dangers of living on barrier islands. The 1900 Galveston Hurricane was America's greatest natural disaster. More than 6,000 people died in this era when the first warnings of a hurricane were its first winds. After the 1900 storm, Galvestonians had a choice of either retreating from the forces of nature or of confronting them once again. For better or worse they chose to stay, and as a result Galveston has the largest protective seawall built on any barrier island in the world. Even if the seawall does successfully protect Galveston against the forces of the sea (not likely in a major hurricane), a heavy environmental price must be paid. Seawalls destroy the beach in front of them, provided they aren't destroyed first. This has happened on South Padre Island, Texas.

There are other problems in paradise. Jetties have been built at the entrance to important harbors in Texas. The jetties make shipping and commerce safer and more efficient, but they also trap sand that would normally feed beaches. The loss of sand translates into an increased danger to beach houses and an increased need for more seawalls. This in turn translates into an increased loss of recreational beaches.

The Texas coastal quandary is repeated in one form or another along much of America's shore. The good news is that, relative to other coastal areas such as Florida and New Jersey, development along the Texas coast is still relatively light. There are still opportunities to learn from the mistakes of others. There are still opportunities for safe and environmentally sound development.

The present volume is the second book in a series being published by the Duke University Press. The series will eventually cover all coastal states. The first volume, entitled *From Currituck to Calabash: Living with North Carolina's Barrier Islands*, is concerned with the barrier-island coast of North Carolina. The success of this book in promoting safe and sound use of the North Carolina islands led to support from federal agencies to produce the other books. Most of the state books will be closely patterned after *From Currituck to Calabash*. Several sections, such as the ones on safe construction and the philosophy of shoreline conservation, are repeated here essentially verbatim. With the use of this book we hope to aid Texas citizens in evaluating the safety and longevity of various portions of their shore. We don't want anyone to be in the frustrating and even tragic position of saying "How was I to know that. . . ."

As part of this coastal safety series Van Nostrand-Reinhold will publish in 1983 *Coastal Design: A Guide for Builders, Planners, and Homeowners* by Orrin H. Pilkey, Sr., Walter D. Pilkey, Orrin H. Pilkey, Jr., and William J. Neal. This book emphasizes coastal construction principles to a much greater extent than the individual state books. We recommend that the prudent coastal citizen also obtain this book.

When we began to plan the Texas volume, there was no question who was best qualified to head up the project. Dr. Robert Morton, in his position as geologist with the Bureau of Economic Geology, University of Texas at Austin, has studied the Texas coast for years and has authored many scientific publications on the subject. Thanks in part to his work, the Texas coast is as well known as any coastal segment in the United States. To our delight, he readily accepted the onerous task of compiling all of the data yet published on Texas islands and beaches, reducing it to a reasonable amount, and then summarizing it in layman's language.

The overall project of producing these books is an outgrowth of initial support from the National Oceanic and Atmospheric Administration through the Office of Coastal Zone Management. The project was administered through the North Carolina Sea Grant Program. We have recently received support from the Federal Emergency Management Agency to expand the book project to all coastal states. The technical conclusions presented herein are those of the authors and do not necessarily represent those of the supporting agencies.

We owe a debt of gratitude to many individuals for support, ideas, encouragement, and information. Peter Chenery of the North Carolina Science and Technology Research Center and Richard Foster of the Federal Coastal Zone Management Agency gave us encouragement and support at critical junctions of this project. Doris Schroeder has helped us in many ways as Jill-of-all-trades over a time span of more than a decade. Doris compiled the index for this volume. Mike Robinson of the Federal Emergency Management Agency worked hard to help us chart a course through the shifting channels of the federal bureaucracy. Dennis Carroll, Jim Collins, Jet Battley, Peter Gibson, Gloria Jimenez, Melita Rodeck, Richard Krimm, Chris Makris, and many others also helped us through the Washington maze. Marcia Franklin typed much of the manuscript and Jerry McAtee, formerly of the Texas General Land Office, reviewed chapter 5. Dan Scranton supervised the illustrations prepared by the Bureau of Economic Geology, University of Texas at Austin. Clint Myers and Barbara Gruver did the maps and prepared the figures, and Ernie Estes provided the cover photograph. We are in the debt of many coastal residents, fellow geologists, coastal engineers, and state and local government officials too numerous to name who enthusiastically provided us with a wealth of data, ideas, and "war stories."

Orrin H. Pilkey, Jr. / *William J. Neal* / series editors

Living with the Texas shore

1. A coastal perspective

In recent years the Texas coast has received considerable attention. More and more people, attracted by its natural beauty and its nearness to major population centers, have been visiting it. And the more that have visited, the more that have wanted to stay. The pressure for new homes and new businesses is growing.

There are good reasons to question the increased development of the Texas coast. For one thing, it is a threat to the environment. For another, building and living on any coast is a risk, and Texas is no exception. It is unreasonable, however, to believe that development can be stopped. Some people will always be willing to risk the loss of life or personal property in order to live near the beach.

Too many people, though, are taking these risks without knowing the facts. This is unnecessary and dangerous. Much is known today about the coast. Unfortunately, a great deal of this information is in technical reports that have not been widely disseminated to the public.

The purpose of this book is to provide people with an inexpensive and readable source of information about the Texas coast that can be used as a guide when buying or building near the shore. The book should also be of interest to beach visitors and present coastal residents. In the pages that follow we briefly describe the Texas coast and provide a short history of its economic development (chapter 1), discuss the geologic processes that continuously modify the beach and adjacent environs (chapter 2), consider the environmental impacts of shoreline engineering (chapter 3), and apply the preceeding information to each segment of the Texas shore (chapter 4). We also summarize the federal and state legislation applicable to the Texas coast (chapter 5) and present simple but cost-effective construction techniques that will improve a building's chances of surviving on the coast (chapter 6). In addition, we provide four appendices: a hurricane checklist, a list of government agencies that supply information on coastal regulations, a list of useful references, and a guide for two field trips.

We hope this book will help make it possible for Texas to avoid many of the problems that other states have experienced in developing their shorelines.

A brief description of the Texas coast

The Texas coast is a barrier coast. It is called this because along most of its 367 miles there are long, narrow strips of sand that parallel the mainland shore. Some of these are joined to the mainland at one point and are thus peninsulas. Others are islands, completely surrounded by water. Between the barriers and the mainland are shallow bodies of water called bays or lagoons. On the other side of them is the Gulf of Mexico (fig. 1.1).

Except for three areas where the mainland is directly exposed

4 Living with the Texas shore

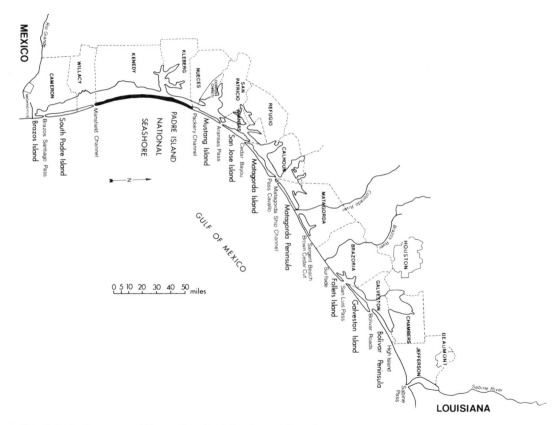

Fig. 1.1. Index map of Texas barrier islands and beaches.

to the Gulf, the islands and peninsulas form an almost continuous barrier along the coast, shielding the mainland against Gulf waves and storms. The points at which the barrier is broken are called *passes* in Texas (in other states they are called *inlets*). These are openings where water flows in and out of the bays and lagoons as the tide changes.

One interesting aspect of Texas's being a barrier coast is that along most of its coast there is not one shore but three: the shore on the mainland, the shore on the bayside of the peninsulas and islands, and the shore on the Gulf side of the peninsulas and islands.

The coast is also characterized by a variety of environments. For example, there are many shallow lakes surrounded by vast marshes on the upper coast between Sabine Pass and High Island (fig. 1.1). This is an area where the mainland is exposed directly to the Gulf. It was formed by deposits that accumulated at the mouths of ancient rivers. Dunes in this area are sparse or absent because of the high rainfall and paucity of sand.

Dunes are slightly better developed and sand is more abundant along the grass-covered barrier islands and peninsulas that form the shoreline between High Island and Surfside. Farther to the west is another area where the mainland directly faces the Gulf. This area formed by the Brazos and Colorado rivers occurs between Surfside and Brown Cedar Cut. Here, as on the upper coast, dunes are sparse or absent and marshes dominate the landscape.

The remainder of the coast is composed predominantly of barrier islands and peninsulas that differ markedly in vegetative cover and dune activity as the average annual rainfall decreases southward. From Brown Cedar Cut to central Padre Island (fig. 1.1) the barriers are covered with a dense blanket of grasses, and the well-developed dunes facing the Gulf are stable except during droughts. In contrast, vegetation is sparse and active dune fields (dunes that are moving) are common on the barriers within and south of the National Seashore.

Farther south, the short segment of shoreline between Brazos Island and Mexico is another area exposed directly to the Gulf. Unlike the other two mainland areas, this one, which was formed by the Rio Grande, is nearly barren. Broad, salt-encrusted sand flats cover its surface. The lack of marshes in this area is explained by the low rainfall and the high salinity of the soil.

The bays and lagoons along the coast are primarily salt-water environments because they are connected to the Gulf by passes. They do, however, receive some fresh water that runs off of adjacent upland areas and that flows in from rivers and streams. In the area within and south of the National Seashore, the bays and lagoons are frequently hypersaline because of the low rainfall and lack of passes.

The above discussion only hints at the diversity and beauty of the Texas coast, qualities that explain the attention it has been receiving lately. What is more difficult to explain is the relatively small amount of attention it has received throughout much of its history.

Key events in the history of the coast

The first settlers

The first coastal dwellers of historical record in Texas were the Karankawas, a nomadic tribe of Indians who hunted game on the barrier islands and fished in the adjacent bays and lagoons. The wisdom of the Karankawas is demonstrated by their choice of campsites. They were not located on the barriers but rather on the mainland where higher elevations provided safety and protection from the devastating forces of nature.

In the sixteenth century Spanish and French adventurers explored the coastal bays and rivers seeking new wealth and establishing new trade routes. The shallow, protected bays also served as hideaways for privateers who were attracted by the precious metals and other valuable cargo transported between Mexico and Spain. Legend has it that pirates lured some vessels aground by walking donkeys along the beach with lanterns tied to their backs. Unsuspecting sailors thought the swaying lights were on the masts of nearby ships.

The story is also told that 300 shipwrecked Spaniards were killed by the Karankawas on a forced march down Padre Island. According to this legend there was only one survivor, who escaped by burying himself in the sand and later fleeing into Mexico.

The lack of fresh water, food, shelter, and adequate building materials on the barriers contributed to the failure of early colonization attempts by the Spanish. The Spaniards' interest in establishing coastal settlements, however, was renewed after LaSalle built Fort Louis on Matagorda Bay around 1685.

The use of barriers for cattle grazing, a practice that continues today along the central coast, probably began in the mid-1700s after Spanish settlements were established near the Rio Grande. The living quarters, outbuildings, and corrals used for ranching represent some of the earliest construction on the barriers.

Early ports and forts

Except for Galveston, which was incorporated in 1838, the earliest commercial developments on Texas barriers other than ranches were quarantine stations, supply docks, or trading posts at entrances to the shallow bays. Of those early settlements, only Port Bolivar, Galveston, and Port Aransas remain today as testimony to the wisdom of early developers. Other ports, however, were not so fortunate. The town of Indianola on Lavaca Bay was a major seaport beginning in 1844, but its history was short-lived. Although it survived yellow fever epidemics and battles of the Civil War, Indianola was demolished by back-to-back hurricanes in 1885 and 1886, and the town never recovered. Now a statue of LaSalle is all that remains to remind us of the bustling port and thriving community that existed only a century ago.

Systematic surveys of the coast were initiated in the early 1850s, shortly after Texas gained statehood. By this time the coastal economy largely depended on shipping, and safe navigation for sailing vessels was important. In order to maintain safe passages and eliminate dangerous shoals and shifting bars, channels were dredged and jetties were built at major passes on the upper coast (Sabine Pass, Bolivar Roads).

By the turn of the century, channels had also been stabilized and enlarged at Aransas Pass, Brazos Santiago Pass, and at the mouth of the Brazos River (Freeport Harbor). For reasons we discuss in chapter 3, these same harbor modifications are contributing to shoreline erosion by further reducing the volume of sand supplied to adjacent beaches.

Beginning with the Mexican War, the barriers played an important role in many military operations and training maneuvers. Their spaciousness, isolation, and strategic location have made them ideal for such use. Military use peaked during World War II when central-coast barriers were the sites of active air bases and bombing ranges, and Galveston Island supported naval observation towers and artillery installations. Most of these facilities were abandoned shortly after the war, leaving Matagorda Air Force Base (now also abandoned) the only active military operation in the 1970s.

Resort communities

The first permanent island dwellers were skilled in survival and lived in harmony with their island. Buildings were placed on the bay side of the islands or on high ground behind the dunes, which provided some protection against wind and flood. Even the best available locations and construction practices, however, were no guarantee that the structures would survive. For example, the storm of 1868 swept away part of Fort Esperanzas, which was built on a relatively safe site on Matagorda Island. Early islanders were constantly confronted by the forces of nature. Commonly these forces were underestimated, and this led to disastrous results. A case in point is the migration of Aransas Pass, which caused the failure of numerous structures and left the lighthouse on Harbor Island stranded more than a mile from the entrance of the pass.

Early resort development of the barrier islands was slow. Access was limited, and social and political pressure to build bridges to the islands was not strong. Thus, few people were able to vacation at the shore.

Development of the Texas barriers didn't begin in earnest until the late 1920s and early 1930s when wooden causeways spanning Laguna Madre connected North and South Padre Islands with the mainland. Even then, resort development was slowed because of the depression and later because of World War II. The first big building boom came in the mid-1950s after permanent causeways were built to North and South Padre. In the 1970s a second wave of resort development was concentrated in those two areas and on Galveston Island and Mustang Island. During this period, beach-cottage development flourished on Bolivar Peninsula. Although smaller communities have formed elsewhere, development has been concentrated in these same areas for the past decade.

High-density development

The high-density development phase of the shore began after World War II (fig. 1.2). Before 1950 and the coming of the bull-

Fig. 1.2. High-density development, South Padre Island.

Fig. 1.3. Undeveloped barrier segment, Mustang Island.

dozer, only a few resorts such as Galveston Island and Port Aransas were in existence; the rest of the islands were either untouched or hardly touched by man (fig. 1.3). The effects of overdevelopment are not yet as apparent in Texas as they are in some other states. Today, even the most casual observer can see the results of overdevelopment along the New Jersey coast. A trip there would be instructive for every Texan; the sight of the New Jersey shore conveys a more dramatic message than the pages of any book (fig. 1.4).

The problem with overdevelopment is, in part, one of destroyed natural beauty. But beauty is in the eyes of the beholder; some would rather see a hot dog stand on the beach than a dune covered with sea oats. There are, however, nonaesthetic problems that pose a more serious threat to coastal residents such as shoreline

1. A coastal perspective   9

Fig. 1.4. Overdevelopment of the New Jersey coast. Note riprap seawalls, total lack of beach, and rubble from destroyed engineering structures.

erosion (fig. 1.5) and other natural hazards that are discussed in chapter 2.

Hurricane history: a stormy past

The Texas coast has taken more than its share of hurricane winds, waves, and floods—considerably more than neighboring Gulf states such as Alabama and Mississippi—perhaps because of

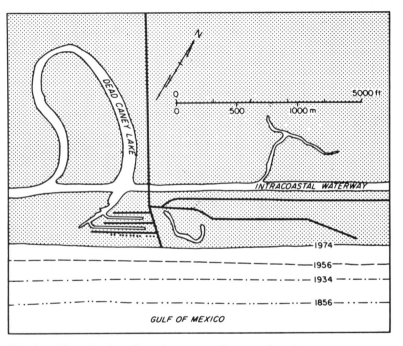

Fig. 1.5. Historic shoreline changes at Sargent Beach.

its position relative to the paths of storms entering the Gulf of Mexico (fig. 1.6).

The effects of seventeenth-, eighteenth-, and even nineteenth-century storms on the Texas coast were generally not well docu-

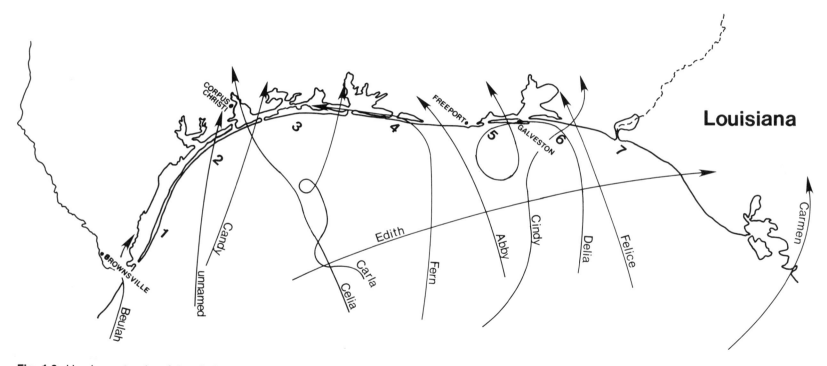

Fig. 1.6. Hurricane tracks of the sixties and seventies.

Table 1.1: Probability of storms (the number expected over a 100-year period)

Sector	Tropical storm	Hurricane	Great hurricane
1	9	8	2
2	12	7	5
3	13	7	4
4	12	9	4
5	18	14	4
6	20	12	4
7	14	8	4

Note: Information modified from Simpson and Lawrence, 1971 (reference 18, appendix C). Sectors 1 through 7 are designated on figure 1.6.

mented because so few people lived on the islands. The fact that so many of the early storms reportedly made landfall near Corpus Christi or Galveston is probably a consequence of the distribution of population centers at that time. The figures vary from one report to another, but at least eighty-five storms have affected this coast since 1880, an average of one tropical cyclone each year. Of these disturbances, a storm of hurricane strength occurred about every 5 to 10 years (table 1.1). Such average figures are misleading. At least five times in the last century as many as three storms have struck the Texas coast in a single year and within a two-month span.

The 1960s brought considerable destruction by hurricanes, whereas the 1950s had brought relative peace and quiet. No one knows what is in store for the 1980s.

Today, anyone on the coast during a hurricane is almost certainly there by choice. In the past, however, people were not warned of a hurricane's approach, and were thus not always able to flee their homes before a hurricane struck. The absence of warning made hurricanes even more feared then than they are today and accentuated the need for safe development.

The decade between 1960 and 1970 was a costly one in terms of hurricane destruction (fig. 1.6). Three names from that period that will long be remembered are Carla, Beulah, and Celia. Each storm was unique in its method of destruction. Nevertheless, property damages from each storm exceeded several hundred million dollars. Total losses from the three storms were on the order of one billion dollars. Hurricane Carla (1961) was characterized by extreme size and great storm surge; Beulah (1967) was known for torrential rainfall and flooding; and Celia (1970) caused extensive damage by extremely high wind speeds. The uncertainty of when and where a storm will strike and what type of storm it will be makes coastal development a risky form of gambling. The conclusion that must be drawn is that any given structure on the coast will experience a major hurricane in its lifetime, perhaps several. When taking out a 25-year mortgage to build or buy in a high-hazard zone, you should pause at length to consider hurricane history.

2. Dynamics of the Texas shore

Except for three stretches, the Texas coast is fronted by narrow, flat ribbons of sand that we refer to as coastal barriers. These barriers, including peninsulas, are very dynamic pieces of land, subject not only to the influence of storms coming in from the ocean but also to less spectacular but no less powerful forces. Anyone living at the Texas coast should be aware of all the forces that are at work. The purpose of this chapter is to acquaint the reader with these forces and to present ways of evaluating them.

Barrier islands: how they were formed

In order to understand the dynamics of the barrier islands, it is necessary to understand how they were formed; this is because the same conditions and processes that formed them continue to affect them today.

The first major factor contributing to the creation of barrier islands is sea level, which has been rising all over the world for thousands of years (fig. 2.1). About 18,000 years ago the Texas shoreline was many miles offshore of its present position on what is now the continental shelf. There were massive glaciers covering the high latitudes of the world and these had bound up a great deal of water. But 18,000 years ago the glaciers started to melt and sea level started to rise.

On the coast the rising water filled valleys, which were formerly the pathways of rivers flowing to the sea, and formed embayments (fig. 2.2). Galveston Bay and Corpus Christi Bay are two prominent examples of these submerged river valleys.

If this were the only process at work, Texas would currently have a very crooked shoreline. Nature abhors a crooked shoreline, however, and will erode and deposit materials in order to straighten it.

Shoreline straightening occurred along the coast as a result of the erosion of the headlands between valleys. The energy of waves striking the headlands moved sand along the beach via currents in the surf zone called longshore currents. Because wave energy and hence current strength were greatest at the headlands, the sand that was being transported couldn't turn the corner and flow into the bay. Instead the sand built up as a spit or sand bar extending into the mouth of the bay (fig. 2.2). Bolivar Peninsula is a good example of such a spit.

Sea level continued to rise, and the land became flooded behind the spits and behind the sand dunes of the old shoreline. As the old dune-beach complexes became detached from the mainland, barrier islands were born (fig. 2.2).

Gently sloping shorelines like those on the Texas coast are the second major factor leading to the creation of barrier islands. Such shorelines can't help themselves; they must have barrier islands in front of them as long as the level of the sea continues to rise. There are barrier islands in front of almost every flat-lying

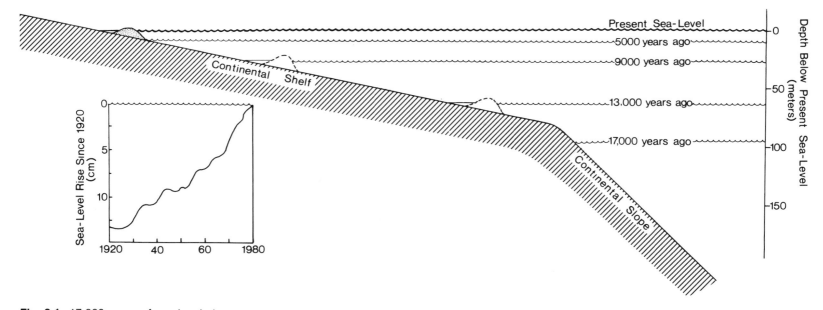

Fig. 2.1. 17,000 years of sea-level change.

coastal plain in the world. In fact, the Texas coast is at one end of an American barrier island coast that stretches 10,000 miles from the south shore of Long Island, down the East Coast, around Florida, and all the way to Mexico.

After the islands formed, they began to move toward the mainland. This is called migration. Needless to say, if the islands were to remain islands, the mainland shore must have retreated too. Viewed in this context, these narrow strips of sand upon which we build our beach cottages are indeed dynamic and ephemeral.

For a long time American barrier islands were moving landward at an impressive rate. Such rapidly moving islands tend to be low and very narrow. Then 5,000 years ago the rise in sea level slowed

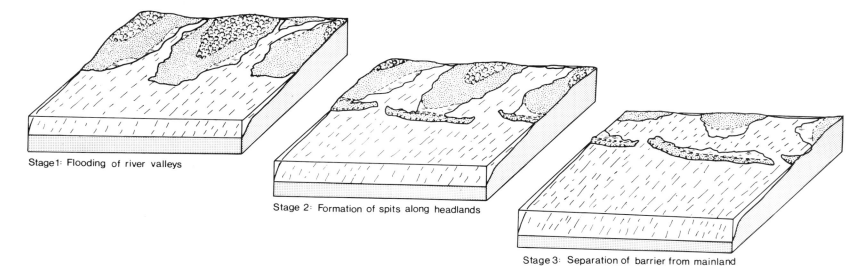

Fig. 2.2. Evolution of a barrier island.

down considerably (fig. 2.1). When the slowdown came, many islands stopped migrating completely and began to grow seaward, or in other words, to widen. Galveston, Matagorda, and San Jose Islands are examples of islands that have widened considerably during the recent pause in sea-level rise.

In the 1930s, however, an immensely important sea-level event occurred. The rise suddenly accelerated. It is now rising at a rate of perhaps slightly more than one foot per century. In order to appreciate what this means to American barrier islands, the second factor affecting barrier islands—that is, the slope of the coast—must be considered. It can easily be demonstrated that for a given rise in sea level a shoreline will move more rapidly across flat-lying land than across steeply sloping land. What this means for American barrier islands, because the slope of much of our coastal

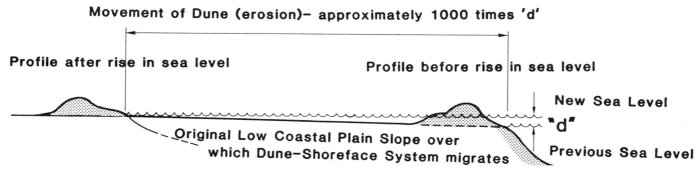

Fig. 2.3. Ratio of horizontal island migration to vertical sea-level rise.

plain is gentle, is that a one-foot rise in sea level should produce an island migration rate of somewhere between 100 and 1,500 feet (fig. 2.3).

More complete figures are available for the rate of shoreline retreat for the Texas coast than for any other stretch of the American shore. The figures show that a sudden and strong acceleration in the rate of shore retreat has occurred in the last 10 to 15 years. If the retreat continues unabated, we can expect the following shoreline recessions during the next century: South Padre, 1,000 to 1,500 feet; Matagorda Peninsula, 500 to 1,000 feet; Mustang Island, 0 to 1,000 feet; Galveston Island, south of the seawall, 1,000 to 2,000 feet; and the Bolivar Peninsula, 500 to 1,000 feet!

Do you want to prove to yourself that islands migrate? If you're standing on one now and it isn't one that has been artificially nourished by hauling in sand, walk to the Gulf beach and look at the seashells. Chances are that on most beaches you will find oyster, clam, or snail shells that once lived behind the barrier island toward the mainland in the lagoon. How did shells from the back side get to the front side? The answer is that the island migrated over the lagoon, and waves attacking and breaking up the old lagoon sands and muds threw the shells up on the present-day beach.

Radiocarbon dates on shells from Atlantic and Gulf Coast beaches frequently reveal ages of thousands of years. In addition, salt-marsh peats that formed in back of islands at some earlier time are exposed occasionally on ocean-side beaches after storms.

On Whale Beach, New Jersey, a patch of mud that appeared on the beach after a storm contained (much to the surprise of some beach strollers) cow hooves and fragments of colonial pottery. The mud was formerly salt marsh on the back side of the island where a colonist had dumped a wagon load of garbage. Since colonial times this particular section of the island had migrated its entire width!

The operation of barrier islands

How islands differ

Every barrier island is unique. Each island evolves by mechanisms that may differ slightly or substantially from those of adjacent islands; thus each island must be understood separately.

For years scientists did not realize this and treated all barriers as if they were the same. Geologists and biologists studying barrier islands in Texas argued with those studying barrier islands in New Jersey. Each group of scientists thought the other group was unobservant. When an investigator attempted to apply what was learned about New Jersey islands to Texas islands, he found the information didn't apply to Texas, and vice versa. Thus, scientists came to realize that there are fundamental differences among look-alike barrier islands.

Let's compare North Carolina and Texas barrier-island systems on a broad scale. If you dam a river in North Carolina, it should not affect the state's barrier islands at all; North Carolina islands get most of their sand from the adjacent continental shelf. Texas islands, however, are nurtured by rivers such as the Rio Grande, the Brazos, and the Colorado which furnish sand directly to the shoreline during every flood. When this supply is stopped by dams, as partially it has been, the beaches begin to starve and retreat more rapidly. Another major difference between Texas and North Carolina barrier islands is in their response to *overwash*. On Texas barrier islands, such as Padre Island and Matagorda Peninsula, overwash passes—where storm waters move across the island— have been flooded again and again during succeeding storms at the same position. On North Carolina islands, the sites of major overwash by storms shift through time. In fact since storm overwash brings sand onto North Carolina islands, previous overwash locations tend to be immune from overwash in the next storm.

Atlantic islands of the southern United States differ from those north of the Mason-Dixon line in the type of vegetation that populates the islands. Paul Godfrey, a botanist at the University of Massachusetts, has observed that the types of vegetation on barriers in the southern United States are responsible for differing degrees of dune development and striking differences in the appearance of islands. For example, the dominant sand-flat plant species is *Spartina patens* on both northern and southern islands, but the northern variety cannot tolerate sand burial whereas the southern variety can. When sand covers a southern island after storm overwash, the grass grows through the new layer of sand within a year. The grass stabilizes the sand and holds it in place. On northern islands, the same overwash will kill the plants, the sand is not stabilized, and it quickly is blown into the surrounding dunes. As a result, the northern islands have more dunes and fewer sand flats than their southern counterparts.

A unique feature of Texas islands has been the historic changes of vegetation patterns. On Padre Island the vegetation pattern seems to be more dynamic than the barrier island itself. Historic vegetation changes can be related to cyclic changes in rainfall. During droughts, dune-stabilizing vegetation dies and dunes are activated until rainfall increases again.

Other natural differences between islands can be the result of such things as average grain size of the sand, island orientation relative to the dominant wind direction, variation in sand supply, amount of shells in the sand, and the character of adjacent tidal passes and shoals. Fine sands retain water better than coarse sands; hence vegetation will restabilize storm-destroyed dunes more rapidly when the sand is fine. Islands where the dominant wind direction is up and down their length tend to have poor dune buildup because not much sand is supplied to the islands from the beach. Islands with a large supply of sand tend to be fatter than those with only a small amount of sand coming ashore. If sand has a high content of shell, as is typical for most barriers in the southern United States, the amount of sand available for dune construction will be reduced. The shelliest beach in Texas is probably Big Shell Beach on the Padre Island National Seashore. Fresh sand that comes ashore during storm washover is winnowed by the wind until a "lag" layer of coarse shells remains. At that point the wind has a tough time getting additional sand because the shells stabilize the sand in much the way vegetation does.

The point we emphasize is that each island has a different story to tell. The island dweller must learn and respond to the unique traits of the particular island he inhabits—if he wants to preserve it. And remember, if you want expert advice, don't ask the old-timer from New Jersey to evaluate your cottage on a Texas island. In fact, don't ask the Padre Island veteran to advise you about Galveston Island.

How islands are similar

Having discussed differences among barrier islands, let's mention some things they have in common. The major mechanisms by which islands move are the same everywhere although the rates and intensities at which these mechanisms operate differ widely.

In order for an island to migrate in Texas, the front (Gulf) side must move landward by erosion and the back (bay) side must do likewise by growth. As it moves, the island must somehow maintain its elevation and bulk.

Front side moves back by erosion. The beach retreats landward because the sea level is rising. This sea-level rise may be the main cause of beach erosion worldwide, although many other local factors such as lack of sand or the effect of man-made structures undoubtedly also cause erosion. The shoreline of the Nile Delta in Egypt, for example, is eroding at an unprecedented rate because the Aswan Dam on the Nile River has cut off the supply of new beach sand. Beaches in California are disappearing for the same reason, that is, dam construction has blocked the flow of river sediments. In many cases, Texas beach erosion is partially a result of sand being trapped by jetties (see chapter 3).

As the sea level rises, the sandy shoreline on the Gulf retreats. People living along the shore call this erosion. The mechanism of shoreline retreat will be discussed later in the chapter. At this point, we need only recognize that the beach retreats horizontally at 100 to 1,500 times the rate of vertical sea-level rise, and that the rate of retreat essentially controls the rate of island migration.

Back side moves back by growth. There are several ways by which an island can widen. One way that islands—especially narrow ones—can be widened is by direct frontal overwash of storm waves from the ocean side of the island. All barrier islands are overwashed to some degree during storms. On large ones (for instance, Mustang and North Padre) the overwash may barely penetrate the first dune line. On low, narrow ones overwash may reach all the way across the island to the bay. Overwashing waves carry sand and deposit it in tongue-shaped or fan-shaped masses called *overwash fans*. When such fans reach into the bay, the island is widened. This process has been going on over the last few years on two narrow Texas barriers: South Padre Island and Matagorda Peninsula.

Overwash is the method of widening used by islands that are in a hurry, that is, those that are migrating rapidly landward. Between 15,000 and 5,000 years ago when the sea level was rebounding rapidly, most American barrier islands were probably of the overwash type.

Today many Texas barrier islands are eroding on both Gulf and bay sides in response to the sea-level rise. These islands are going through the first stage of converting themselves from wide islands to narrow islands. In this way the islands once again become overwash islands that are capable of migrating. For example, as long as Mustang Island is as wide as it is now, overwash from the Gulf side will not transport sand all the way to the bay side. One or two hundred years from now, however, the island may be thin enough for sand to be deposited in the bay as Gulf shoreline erosion continues. This would result in true island migration. If the sea-level rise continues, in a few hundred years most American barrier islands will be totally unlike their present-day ancestors.

Texas islands also widen or build landward through the addition of sand blown into the bay. The best example of this island-widening mechanism is on North Padre Island. Large volumes of sand roll and bounce across the backbarrier flats to eventually be deposited in the bay. Grain by grain the bay is shoaled (fills up with sand) and the shoreline marches into the bay.

The island maintains its elevation during migration. The remaining problem of a migrating island is how to retain its bulk or elevation as it moves toward the mainland. This problem is solved by two processes: dune formation and overwash-fan deposition.

Dunes are formed by the wind from sand blown in from the beach. The sand on the beach may come from nearby rivers or it may have been pushed ashore by waves during good weather. During the storms, sand is transported offshore. Whatever its

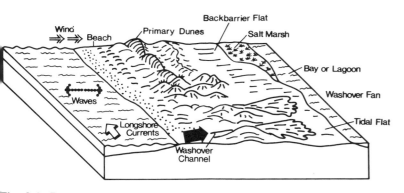

Fig. 2.4. Barrier-island environments.

source, the more sand that comes ashore the larger the dunes and the higher the island. Matagorda Island is higher than the adjacent Matagorda Peninsula because it has a larger sand supply.

Barrier island environments are interrelated

Another important thing that must be understood is that barrier-island environments are interrelated (fig. 2.4). Each environment is part of an overall integrated system and, to some degree, depends on or affects other environments within the system. Specific environments are discussed in chapter 4.

Perhaps the best example of one environment affecting others in the system is provided by the role of the Gulf beach. The beach is important because (1) it alters its shape during storms in such a way as to minimize fundamental damage to the islands by waves, and (2) it is the major source of sand for the entire island or coastal strip. Examples of the ways in which man has interfered in the integrated system may best illustrate these functions.

Dr. Paul Godfrey, who we mentioned earlier, has discovered that the long, continuous, artificial dune built by the National Park Service on the Outer Banks of North Carolina near Cape Hatteras is causing erosion on the *bay side* of the island. The problem is that the artificial dune prevents sand from being washed over the island during storms. Before the dune was built, overwash frequently reached the back side of the island, and a new salt marsh was formed on the edge of the new overwash fan. The newly formed *Spartina* marsh is an excellent erosion buffer against bay-side waves. By preventing overwash, the frontal dune on the ocean side of the island precludes new marsh growth and increases the bay-side erosion rate.

Dune buggies and other off-road vehicles can also damage barrier-island environments. Dune buggies can prevent dunes from stabilizing (becoming stationary), and destabilization (moving sand) may result in destroyed dunes and vegetation, or even in sand-dune migration.

Another common mistake on barrier islands is related to road construction. On several Texas barrier islands if you drive along a road that parallels the beach and if you look toward the beach, you will see notches through the dunes. The notches indicate an access road to the beach, a washover channel, or sometimes just a foot path that has been used for years. Access, however, works two ways and often the access route from land to sea is the access route of the sea to the land in the next storm.

Just as environments on a single island depend on one another, so do environments on adjacent islands. The beaches on our islands are like flowing rivers of sand. Frequently islands depend on neighboring islands for sand supply. When this sediment supply is cut off by inlet dredging or construction of jetties, the erosion rate of the Gulf beaches increases.

Beaches: the dynamic equilibrium

The beach is one of the earth's most dynamic environments. The beach—or zone of active sand movement—is always changing and always migrating, and we now know that it does so in accordance with certain natural laws. The natural laws of the beach formulate a beautiful, logical environment that builds up when the weather is good, and strategically (but only temporarily) retreats when confronted by big storm waves. This system depends on four factors: site of the waves, rate of sea-level rise, amount of beach sand, and the shape of the beach. The relationship among these factors is a natural balance referred to as a "dynamic equilibrium" (fig. 2.5): when one factor changes, the others adjust accordingly to maintain a balance. When man enters the system incorrectly—as he often does—the dynamic equilibrium continues to function in a predictable way but in a way that is harmful to man.

Answers to the following often-asked questions about beaches

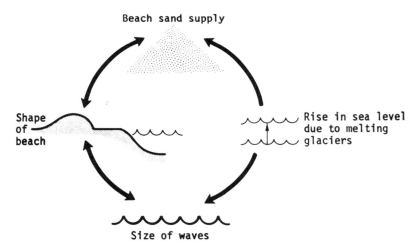

Fig. 2.5. The dynamic equilibrium of the beach.

may clarify the nature of this dynamic equilibrium. It is important to keep in mind that the beach extends from the toe of the dune to an offshore depth of 30 to 40 feet. It is the zone of active sand movement during storms. The part on which we walk is only the upper beach.

How does the beach respond to a storm?

Old-timers and storm survivors from barrier islands have frequently commented on how beautiful, flat, and broad the beach is after a storm. The flat beach can be explained in terms of the dynamic equilibrium: as wave energy (height) increases, materials move to change the shape of the beach. The reason for this storm response is logical. The beach flattens itself in order to make storm waves expend their energy over a broader and more level surface. On a steeper surface, storm-wave energy would be expended on a smaller area, causing greater damage.

Figure 2.6 illustrates the way in which the beach flattens. Waves take sand from the upper beach or the first dune and transport it to the lower beach. If a hot dog stand or beach cottage happens to be located on the first dune, it may disappear along with the dune sands.

The beach can lose a great deal of sand during a storm. Much of it will come back, however, gradually pushed shoreward by fair-weather waves. As the sand returns to the beach, the wind takes over and slowly rebuilds the dunes, storing sand to respond to nature's next storm call. Return of the beach after a storm may take months or even years. In order for the sand to come back, of course, there should be no man-made obstructions—such as a seawall—between the first dune and the beach.

Sometimes besides simply flattening, a storm beach will also develop one or more offshore bars. The bars serve the function of tripping the large waves before they reach the beach. The sand bar produced by storms is easily visible during calm weather as a line of surf a few tens of yards off the beach.

Where does the beach sand come from?

Along most of the Atlantic barrier coast plus Florida's west coast, the sand comes from the adjacent continental shelf. It is pushed up to the beach by fair-weather waves. Additional sand, sometimes very large quantities of it, is carried laterally by longshore currents that move sediment in the surf zone parallel to the beach. Rivers contribute sand directly to barrier beaches only along the Gulf coast, starting with Florida's Appalachicola River. Along the rest of America's barrier chain, sand carried by rivers does not make it to the coast. Rather it is deposited far inland at the heads of estuaries.

It is important for beach dwellers to know or at least to have some appreciation of the source of sand for their beach. If, for example, there is a lot of longshore sand transport in front of your favorite beach, the beach may disappear if someone builds a groin "upstream" or in the direction from which the sand is coming (see chapter 3). In Texas, sand transported by longshore currents is probably the principal source of beach sand today.

Longshore currents are familiar to anyone who has been swimming in the ocean; they are the reason you sometimes end up somewhere down the beach, away from your beach towel. Such currents result from waves approaching the shore at an angle; this causes a portion of the energy of the breaking waves to be directed along the beach. When combined with breaking waves, the weak current is capable of carrying large amounts of sandy material for miles along a beach.

22   Living with the Texas shore

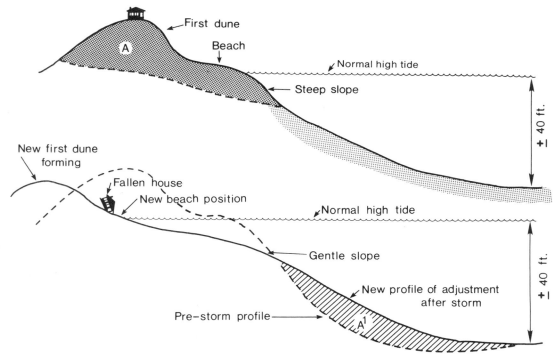

(Shaded area A¹ is approximately equal to shaded area A.)

Fig. 2.6. Beach flattening in response to a storm.

How does the beach widen?

Most Texas beaches are wider during the summer than during the winter. This is because the gentle summer waves have a pushing action on the sea floor and gradually push sand ashore. The reverse happens during the winter season, which produces higher waves, and during storms. Sand is moved seaward and the upper beaches become narrow. If one looks at the entire beach from the high-tide line to water depths of 30 to 40 feet, however, one would find that the winter beach is slightly flatter than the summer beach.

Where do seashells come from?

Surprisingly, the majority of the shells on many barrier island beaches can be called fossils. Many Texas shells have radiocarbon ages that are thousands of years old; they are usually less than 10,000 years old, however.

If you use a shell book (reference 45, appendix C) carefully to identify specimens from a beach, you will find that bay or lagoon shells are very common on the Gulf beach. As the islands migrated landward, they ran over the shells that once lived in back-island environments. In a few hundred or a thousand years, the bay shells were thrust up onto the Gulf-side beach.

As any beach buff knows, however, not all beach seashells are fossils by any means. The coquina clam (*Donax*) lives in the upper beach and hastens to rebury itself when exposed by sand-castle builders!

Why do beaches erode?

As we have already pointed out, "beach erosion" is primarily caused by the sea-level rise—presently judged to be about one foot per century along American shores. We can be thankful in Texas that we don't have the higher rate of sea-level rise experienced along the New England coast. (The reason the sea-level rise can be different in different coastal areas is that the land also may be slowly sinking or rising relative to sea level.)

Working in conjunction with the sea-level rise are the many forces of nature at the shore including waves, tides, and the wind; these are aided and abetted by the effects of man: jetties and groins, dams on rivers, and even perhaps the tracks cut into the sand by beach vehicles.

A geologist once spoke at a luncheon in Virginia Beach, Virginia, and told the audience that the most serious problem facing their eroding shoreline was the rising sea level. A local reporter, mocking the speech, reported in banner headlines that we must "beware the year 4,000 for then our houses will be underwater." The joke is on him for by then his house probably will be 20 miles out at sea as well as in water 30 feet deep! The impact of the sea-level rise will be felt within our own lifetime and should not be regarded as such a long-term event as to be of no consequence. In fact, the National Academy of Science has warned that the rise is expected to accelerate in coming decades.

How can I tell if my shoreline is eroding?

One sure sign of beach erosion on Texas beaches is a lot of shells in the sand. The one exception to this is Big Shell Beach on Padre Island. Another clue is a steep bluff face with plant roots protruding in the dunes adjacent to the beach. Even more obvious clues are a house in the surf at high tide, pavement from a road on the beach, a septic tank on the beach, or a seawall in front of a condominium. Seawalls generally are not built unless erosion is occurring.

The best way to be certain whether erosion is occurring on your beach is to refer to chapter 4 of this book or to some of the publications listed in appendix C.

If most of the Gulf shorelines are eroding, what is the long-range future of Texas development?

The long-range future of beach development in Texas will be a function of how individual island communities are able to respond to their eroding shore. Those communities that choose to protect their frontside houses at all costs need only look to portions of Galveston Island to see the end result. The life span of houses can unquestionably be extended by stabilizing a beach (stopping the erosion). The ultimate cost of stopping erosion on a barrier island is, however, loss of the beach. The time span required for destruction of the beach is highly variable and depends on the island. Usually a barrier-island seawall that is long will do the trick in 10 to 30 years. Sometimes a single storm will permanently remove a beach in front of a seawall.

Except for the Galveston seawall, which is the mightiest seawall on any barrier island in the world, no Texas seawall on the open Gulf has lasted without damage for more than a few years.

If a community can somehow buy, move, or let the front row of buildings fall in as their time comes, the beaches can be saved in the long run. So far in America the primary factor involved in shoreline decisions, one which every beach community must sooner or later confront, has been money. Communities without a large tax base like Surfside let the beach roll on. Richer communities like South Padre Island attempt to stop it. Surfside will have a beach for future generations; South Padre will not.

Are the shorelines on the back sides of Texas islands eroding?

Most are; some are not. In fact, some are building landward. The present condition of bay shorelines has been qualitatively determined (references 22-28, appendix C); however, rates of erosion for most shorelines have not been determined. Published erosion rates for the shorelines of Mustang Island (reference 31, appendix C) and Matagorda Peninsula (reference 68, appendix C) indicate that a recession of between 1 and 5 feet per year is typical.

What can I do about my eroding beach?

This is a complex question and is partially answered in chapter 3. If you are talking about the Gulf shoreline, there is nothing you can do unless (1) you are wealthy or (2) the Corps of Engineers steps in. Your best response, especially from an environmental and economic standpoint, is to move your threatened cottage elsewhere. The bottom line in trying to stop erosion of ocean shoreline is that the methods employed will ultimately increase the erosion rate. For example, bulldozing sand up from the lower beach will steepen the profile and cause the beach to erode more rapidly during the next storm. Pumping in new sand (replenishment) costs a great deal of money (at least one million dollars per mile) and in most cases the artificial beach will disappear much more rapidly than its natural predecessor. Almost no beach replenishment has been carried out in Texas, in many cases because of the lack of suitable sand supplies.

An interesting side note has been added by geologists from the Bureau of Economic Geology. They have noted that when heavily used public beaches are being cleaned, a great deal of sand is removed along with the seaweed, logs, bottles, and tons of plastic trash. One thing an eroding beach does not need is additional loss of sand. Furthermore it is quite possible that the trash on a beach will retard (slightly) the rate of sand loss. Thus the dilemma of balancing the fear of the erosion rate and the desire for a clean beach is presented to some Texas beach communities.

In summation, there are many ways to stop erosion in the short run if lots of money is available, but in the long run (30 to 100 years), erosion cannot be halted except at the cost of losing the beach.

Hurricane origin: a storm is born

Each year hurricane season begins on the first of June and lasts until the end of October. During this period, air movement and water temperatures in the Atlantic Ocean and Gulf of Mexico favor the formation of tropical cyclones, a group of meteorological phenomena that range from weak, disorganized tropical disturbances to intense, well-organized hurricanes.

Heat derived from the ocean is an important element in the formation of tropical cyclones. Evaporation of seawater produces a layer of moist air over the ocean that is separated from an overlying layer of warm, dry air. As the lower layer of moist air warms, it also rises, then cools and condenses. The cooling and condensation processes release heat that is transferred to the surrounding air, causing it to rise. The mass of rising air forms an area of low barometric pressure (tropical disturbance). Such disturbances, which are common in the tropics, may intensify when more warm air moves in to replace the rising air. As the wind patterns become organized, the earth's rotation causes the air currents in the northern hemisphere to flow in a counterclockwise direction, and a spiraling air mass is formed (tropical depression). Air drawn to the middle of the rotating spiral moves upward and produces a chimney effect. This area of rising air and low pressure becomes the so-called eye of the storm.

A tropical depression is upgraded to a tropical storm if the central barometric pressure continues to decrease and the wind attains speeds of 39 miles per hour; when sustained winds reach

74 miles per hour the storm is classified as a hurricane. Hurricanes are further subdivided into five categories (minimal, moderate, extensive, extreme, and catastrophic) depending on the surge height, wind velocities, central pressure, and amount of damage. Hurricane Carla, which crossed the Texas coast in 1961, was an extreme hurricane whereas Camille, which hit Mississippi in 1969, was a catastrophic hurricane.

Hurricane forces: wind, waves, and washover

Once a hurricane forms or enters the Gulf of Mexico it usually moves in a northerly or a northwesterly direction (see fig. 1.6). As the storm moves across the Gulf it may gain strength and move rapidly onshore or it may stall, lose strength, and dramatically change direction. Such erratic and unpredictable behavior makes it difficult to forecast the time and site of a landfall.

The most destructive storms are those that develop extreme wind velocities (175 to 200 mph), deep-water wave heights (40 to 50 feet), and storm-surge elevations (15 to 20 feet). These three forces in conjunction with intensive rainfall and locally spawned tornadoes cause widespread damage and economic losses on the order of hundreds of millions of dollars. The physical damage is usually concentrated near the eye because the forces of the storm are greatest about its center. Wind velocities and flood heights are greatest to the right of the storm when viewed looking toward the land. The counterclockwise circulation of air generates onshore or southwesterly winds as the storm approaches the coast and off-shore or northeasterly winds shortly after the eye crosses the coast. The wind piles water up along the coast and produces strong longshore currents that run parallel to the coast; the wind also forms steep waves that erode the beach and dunes. The strong winds and low barometric pressure create unusually high water levels in the Gulf known as the storm surge. The surge accompanies the storm and causes a rapid rise in water that lasts several hours. Surge heights at the Gulf shoreline may be 10 to 20 feet above sea level (table 2.1), and surge heights in the bays may even be greater. During Hurricane Carla, water depths were 22 feet above mean sea level at some locations in Lavaca Bay. Extensive areas of the low-lying, poorly drained coastal plain are inundated by the storm surge combined with torrential rainfall and stream flooding (table 2.2).

Storm damage is usually severe where surge heights exceed the dune elevations. Those areas are washed over by the wind-driven

Table 2.1: Still-water surge levels for twentieth-century hurricanes

Beach segment	Surge height	Year
Sabine Pass to Galveston	7	1942
Galveston to Freeport	20	1900
Freeport to Colorado River	11	1949
Colorado River to Port Aransas	12	1961
Port Aransas to Central Padre	9	1970
Central Padre to Rio Grande	13	1933

Notes: Elevations are given in feet above mean sea level. Information from the National Hurricane Center.

Table 2.2: Summary of natural hazards and their impact on the Texas coast

Number of hurricane landfalls, 1900-1982	29
Area (square miles) of salt-water flooding, Hurricanes Carla and Beulah	3,164
Area (square miles) of fresh-water flooding, Hurricane Beulah	2,187
Area (square miles) of fresh-water flooding by hurricane rainfall (floodplains), northern part of Coastal Zone only	2,073
Area (square miles) below elevation of 20 feet (MSL): subject to salt-water flooding by tidal surge	5,787
Number of active or potential hurricane washover channels	137
Number of miles of Gulf beach erosion greater than 10 feet per year (long term)	47
Number of miles of Gulf beach erosion: from 5 to 10 feet per year (long term)	50
Number of miles of Gulf beach erosion: from 0 to 5 feet per year (long term)	104
Number of miles of bay and lagoon shoreline erosion	403
Area (square miles) of land subsidence: greater than 5 feet	227
Area (square miles) of land subsidence: from 1 to 5 feet	1,080
Area (square miles) of land subsidence: from 0.2 to 1 foot	5,422
Number of miles of known active surface faults	96
Number of miles of Gulf shoreline	367
Number of miles of bay-lagoon shoreline	1,100
Area (square miles) of bays and lagoons	2,075

Note: This table is reproduced from Brown and others, 1974 (reference 57, appendix C).

Fig. 2.7. Washover channels cut through Matagorda Peninsula by Hurricane Carla in 1961. Source: National Oceanic and Atmospheric Administration.

surge and receive the full force of storm waves. Washover processes encompass both erosion and deposition at the peak of the storm. Some of the sand eroded from the beach and dunes is carried landward by strong currents and breaking waves. The sand is deposited as a blanket when sheetwash occurs and as individual fans when flow is confined to washover channels eroded across the island. Washover channels are extremely dangerous building sites because they attract high-velocity currents in subsequent storms. Some washover channels completely cut through the island, causing temporary isolation of large barrier segments (fig. 2.7).

Together, the wind, waves, and washover smash buildings, cause power failures, pollute water supplies, disrupt sewage services, and block transportation routes. A knowledge of previous storm characteristics (wind velocities, surge heights, amounts of rainfall, distances of beach and dune erosion, economic losses) and local ground conditions (surface elevations, soil types, drainage paths, evacuation routes) will assist greatly in an individual's selecting a "safe" homesite on the Texas coast.

3. Man and the shoreline

Stabilizing the unstable

Shoreline engineering is a general phrase that refers to any method of changing or altering the natural shoreline system in order to stabilize it. Methods of stabilizing shorelines range from the simple planting of dune grass to the complex emplacement of large seawalls using draglines, cranes, and bulldozers. The benefits of such methods are usually shortlived. Locally, shoreline engineering may actually cause shoreline retreat, as evidenced by the beach in front of the Galveston seawall. Beach erosion caused by man may be greater and more spectacular than nature's own.

The Gulf beaches of Texas have fewer structural failures (broken seawalls and bulkheads) on them than do other coastal states (New Jersey, for example), mainly because fewer projects have been carried out. But the percentage of structural failures is just as high in Texas as elsewhere. Seawalls on South Padre, North Padre, and Sargent Beach have all failed or been severely damaged in storms. Only the massive Galveston wall has remained intact over a long period of time, but in some portions its footings are threatened by erosion.

Experience elsewhere has shown that the economic and environmental price for shoreline stabilization is high indeed. Making the public aware of how high is one of the purposes of this book. There are, of course, situations in which stabilization is an economic necessity. Channels leading to our state ports at Houston, Freeport, Corpus Christi, and Brownsville, for example, must be maintained.

There are three major ways by which shorelines are stabilized. These methods are listed below, in decreasing order of environmental safety.

Beach replenishment

If you must repair a beach, this is probably the gentlest method. Replenishment consists of placing sand on the beach and building up the former dunes and upper beach. Sufficient money is never available to replenish the entire beach out to a depth of 30 to 40 feet. Thus, only the upper beach is covered with new sand, so that in effect, a steep beach is created (fig. 3.1). This new steepened profile often increases the rate of erosion. Few studies have been made of this, but replenished beaches often disappear faster than the natural beaches they replaced. In beach replenishment, sand is either pumped from the adjacent continental shelf, a pit on the land, or the offshore bars, but most often from the adjacent bay or lagoon. Bay sand, however, tends to be too muddy and too fine; it quickly washes off the beach. Furthermore, dredging in the bay can disturb the ecosystem, and the hole created in the bay affects waves and currents, sometimes harming the back side of the island. Perhaps the best source of sand environmentally, but the

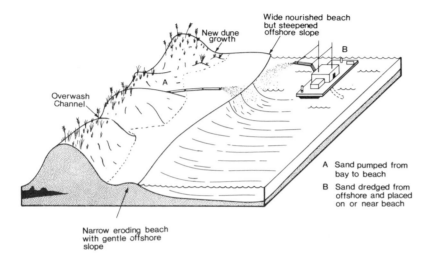

Fig. 3.1. Beach nourishment.

most costly, is the continental shelf; however, Dr. Victor Goldsmith of the University of South Carolina warns that when a hole is dug on the shelf for replenishment sand, wave patterns on the adjacent shoreline will likely be affected. Off the Connecticut coast, wave patterns changed by a dredged hole on the shelf quickly caused the replenished beach to disappear.

Very little beach replenishment has been carried out along the Texas coast. Partly this is because a suitable supply of offshore or bay sand is missing from the Texas areas where the sand is needed. Another factor is that the communities in need (for instance, Surfside and Sargent Beach) have not been able to pay for their share of a Corps of Engineers replenishment project.

One replenishment project on a bay shoreline, the North Beach of Corpus Christi, illustrates an important principle of beaches and beach replenishment. The sand utilized to replenish this beach was obtained from a nearby river. The size of the grains of river sand was much larger than the size of the grains of beach sand. The slope of a beach is controlled in part by grain size; coarse grain sizes form steep slopes and fine grain sizes form gentle slopes. The coarse river sand changed the slope of the Corpus Christi beach making it less suitable than before for swimming and recreation for small children.

Beach replenishment, then, upsets the natural system; it is costly and temporary, requiring subsequent replenishment projects in order to remain effective. The U.S. Army Corps of Engineers, in fact, refers to beach replenishment as an "ongoing" project. Nevertheless, it is usually less harmful to the total dynamic equilibrium than the structural methods of stabilization such as groins and seawalls.

Groins and jetties

Groins and jetties are walls built perpendicular to the shoreline. A jetty, often very long (sometimes miles), is intended to keep sand from flowing into a ship channel. Groins, much smaller walls built on straight stretches of beach away from channels and inlets, are intended to trap sand flowing in the longshore (surf-zone) current.

Both groins and jetties are very successful sand traps. If a groin is working correctly, more sand should be piled up on one side of it than on the other. The problem with groins is that they trap sand that is probably flowing to neighboring beaches. Thus, if a groin on one beach is functioning well, it must be causing erosion elsewhere by starving another beach (fig. 3.2).

Good examples of groin construction in Texas are found in front of the Galveston seawall. Since most of the sand that is lost at this location is probably taken by storm waves moving sand straight offshore, it is not likely that the groins have had any important effect on the shoreline. However, the sand trapped by these groins protects the toe of the seawall from erosion.

Jetties can be found in abundance along the Texas coast. They are designed to allow safe entrance into Texas ports of ships of much deeper draft than the natural channels would have allowed. Texas jetties have been quite successful in their mission of enhancing safe entry and exit from harbors. Jetty construction in other states has not always been so fortunate (for instance the Barnegat Jetties in New Jersey). Residents of the Texas shore, however, are paying a price for the success of their jetties. It has been estimated that as much as 50 percent of the sand supply for Texas beaches is trapped by jetties. What this means in the long run is that the jetties will add measurably to the rate of the Texas shoreline recession.

Seawalls

Seawalls, built back from and parallel to the shoreline, are designed to receive the full impact of the waves during storms. Present in almost every highly developed coastal area, seawalls are not yet common on the Texas coast. A more common type of structure is the bulkhead, a type of seawall placed farther from the shoreline in front of the first dune—or what *was* the first dune—and designed to take the impact of occasional storm waves only.

Building a seawall or bulkhead is a very drastic measure on Gulf beaches because it causes damage in the following ways:

1. It reflects wave energy, ultimately removing the beach and steepening the offshore profile. The length of time required for beach loss to occur is 1 to 50 years. The steepened offshore profile increases the storm-wave energy striking the shoreline, which increases erosion.
2. It increases the intensity of longshore currents, hastening removal of the beach (fig. 3.3).
3. It prevents the exchange of sand between dunes and beach. Thus, the beach cannot supply new sand to the dunes on shore.
4. It prevents storm waves from removing sand from the first row of dunes, so the beach can flatten. Thus, it prevents beaches from responding naturally to a storm (see chapter 2).

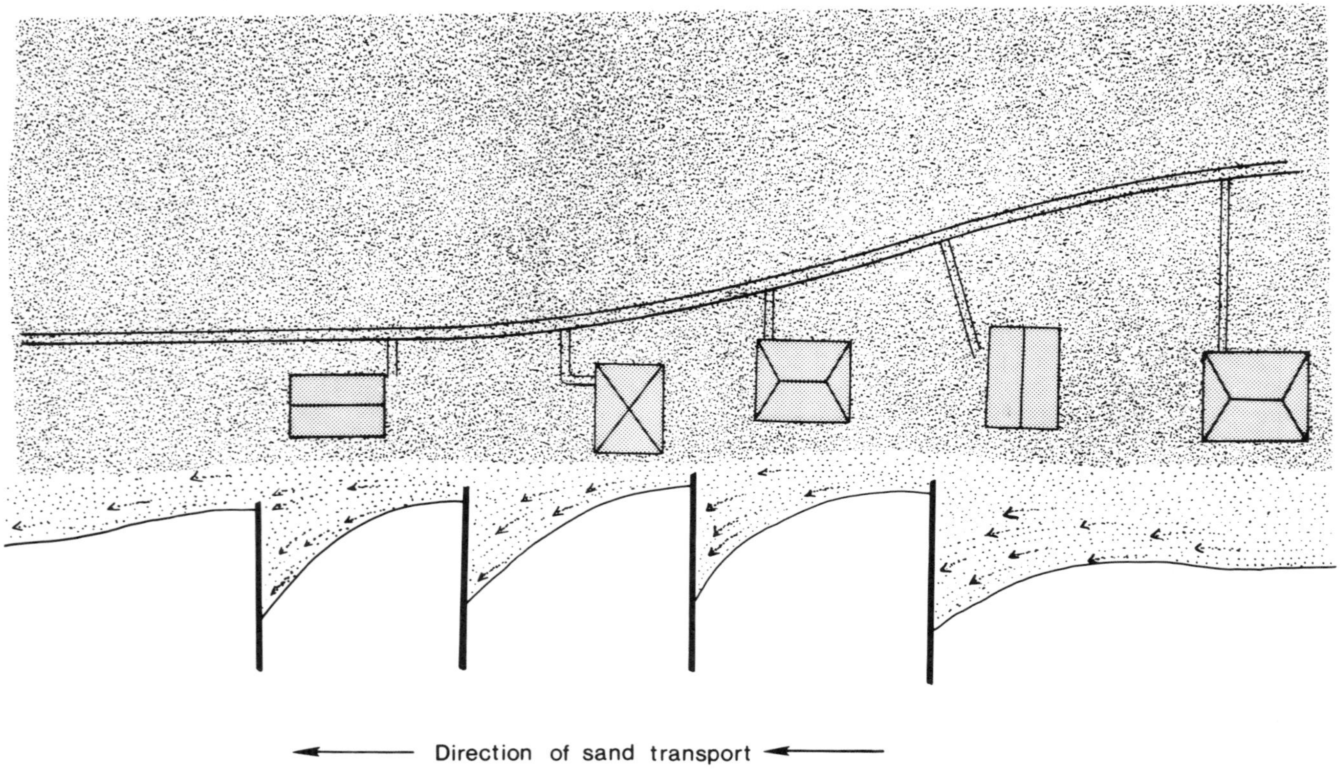

Fig. 3.2. Groined shoreline.

3. Man and the shoreline   33

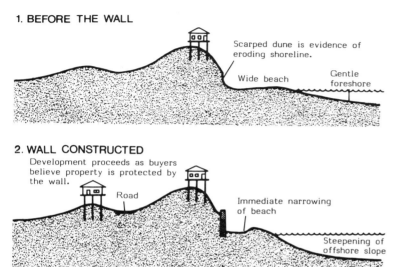
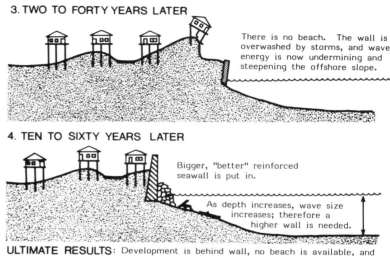

Fig. 3.3. Saga of a seawall.

5. It concentrates wave and current energy at the ends of the wall, increasing erosion at these points.

The effects described here can be seen in front of the seawall at Galveston. Here the natural shoreline profile has retreated well beyond the ends of the seawall. Each year the wall and the buildings it protects will be sticking a little farther out on the beach. The beach is steeper and narrower in front of the seawall than it is to the north or south. Careful observation will also reveal that the big waves are breaking closer to the beach in front of the wall than elsewhere. The emplacement of a seawall is an irreversible act with limited benefits. By gradually removing the beach in front of it, most privately erected seawalls must eventually be replaced with bigger and more expensive ones.

The long-range effect of seawalls can be seen in New Jersey,

Galveston, and Miami Beach. In Monmouth Beach, New Jersey, the town building inspector told of the town's seawall history. Pointing to a seawall he said, "There were once houses and even farms in front of that wall. First we built small seawalls and they were destroyed by the storms that seemed to get bigger and bigger. Now we have come to this huge wall which we hope will hold." The wall he spoke of, adjacent to the highway, was high enough to prevent even a glimpse of the sea beyond. There was no beach in front of it, only remains of old seawalls, groins, and bulkheads for hundreds of yards of sea.

A philosophy of shoreline conservation: "We have met the enemy and he is us"

In 1801, Postmaster Ellis Hughes of Cape May, New Jersey, placed the following advertisement in the *Philadelphia Aurora*:

> The subscriber has prepared himself for entertaining company who uses sea bathing and he is accommodated with extensive house room with fish, oysters, crabs, and good liquors. Care will be taken of gentlemen's horses. Carriages may be driven along the margin of the ocean for miles and the wheels will scarcely make an impression upon the sand. The slope of the shore is so regular that persons may wade a great distance. It is the most delightful spot that citizens can go in the hot season.

This was the first beach advertisement in America and sparked the beginning of the American rush to the shore.

In the next 75 years, six presidents vacationed at Cape May. At the time of the Civil War it was certainly the country's most prestigious beach resort. The resort's prestige continued into the twentieth century. In 1908, Henry Ford raced his newest model cars on Cape May beaches.

Today, Cape May is no longer found on anyone's list of great beach resorts. The problem is not that the resort is too old-fashioned but that no beach remains on the cape (fig. 3.4).

The following excerpts are quoted from a grant application to the federal government from Cape May City. It was written by city officials in an attempt to get funds to build groins to "save the beaches." Though it is possible that its pessimistic tone was exaggerated to enhance the chances of receiving funds, its point was clear:

> Our community is nearly financially insolvent. The economic consequences for beach erosion are depriving all our people of much needed municipal services. . . . The residents of one area of town, Frog Hollow, live in constant fear. The Frog Hollow area is a 12 block segment of the town which becomes submerged when the tide is merely 1 to 2 feet above normal. The principal reason is that there is no beach fronting on this area. . . . Maps show that blocks have been lost, a boardwalk that has been lost. . . . The stone wall, one mile long, which we erected along the ocean front only five years ago has

Fig. 3.4. Cape May, New Jersey, seawall (1976).

already begun to crumble from the pounding of the waves since there is little or no beach. . . . We have finally reached a point where we no longer have beaches to erode.

Texas will not have to wait a century and a half for this crisis to reach its shores. The pressure to develop is here and increasing. Like the original Cape May resort, many of our structures are not placed far back from the shore; nor have we been so prudent as to place all structures behind dunes or on high ground. Consequently, our coastal development is no less vulnerable to the rising sea than was Cape May's, and no shoreline engineering device will prevent its ultimate destruction. The solution lies in recognizing certain truths about the shoreline.

Truths of the shoreline

Cape May is the country's oldest shoreline resort. Built on a shoreline that migrates, it is a classic example of the poorly developed American shoreline, and one from which Texas can learn. From examples of Cape May and other shoreline areas, certain generalizations or "universal truths" about the shoreline emerge quite clearly. These truths are equally evident to scientists who have studied the shoreline and old-timers who have lived there all of their lives. As aids to safe and aesthetically pleasing shoreline development, they should be the fundamental basis of planning on any barrier island.

There is no erosion problem until a structure is built on a shoreline. Beach erosion is a common, expected event, not a natural disaster. Shoreline erosion in its natural state is not a threat to the coast. It is, in fact, an integral part of coastal evolution (see chapter 2) and the entire dynamic system. When a beach retreats it does not mean that it is disappearing; it is migrating. Many developed shorelines especially on barrier islands are migrating at surprisingly rapid rates, though only the few investigators who pore over aerial photographs are aware of it. Whether the beach is growing or shrinking does not concern the visiting swimmer, surfer, hiker, or fisherman. It is when man builds a "permanent" structure in this zone of change that a problem develops.

Construction by man on the shoreline causes shoreline changes. The sandy beach exists in a delicate balance with sand supply, beach shape, wave energy, and sea-level rise. This is the dynamic equilibrium discussed in chapter 2. Most construction on or near the shoreline changes this balance and reduces the natural flexibility of the beach (fig. 3.5). The result is change which often threatens man-made structures. Dune-removal, which often precedes construction, reduces the sand supply used by the beach to adjust its profile during storms. Beach cottages—even those on stilts—may obstruct the normal sand exchange between the beach and the shelf during storms. Similarly, engineering devices interrupt or modify the natural cycle (fig. 3.6; see also figs. 2.5 and 2.6).

Fig. 3.5. Galveston seawall. Note the absence of a beach.

Shoreline engineering protects the interests of a very few, often at a very high cost in federal and state dollars. Shoreline engineering is carried out to save beach property, not the beach itself. Shore-stabilization projects are in the interest of the minority of beach property owners rather than the public. If the shoreline were allowed to migrate naturally over and past the cottages and hot dog stands, the fisherman and swimmer would not suffer. Yet beach property owners apply pressure for the spending of tax economic justification for extensive and continuous shoreline-stabilization projects. For example, to spend tax money for replenishing Coney Island, New York, which accommodates tens of thousands of people daily during the summer months is more justifiable than to spend tax dollars to replenish a beach that serves only a small number of private cottages. In the case of the former, the beach maintenance is in the interest of the public that pays for it, whereas in the latter case the expenditure amounts to middle-class welfare.

Shoreline engineering destroys the beach it was intended to save. If this sounds incredible to you, drive to New Jersey and examine their shores. See the miles of "well protected" shoreline—without beaches (fig. 3.4)!

The cost of saving beach property through shoreline engineering is usually greater than the value of the property to be saved. Price estimates are often unrealistically low in the long run for a variety of reasons. Maintenance, repairs, and replacement costs are typically underestimated, because it is erroneously assumed that the big storm, capable of removing an entire beach-replenishment project overnight, will somehow bypass the area. The inevitable hurricane, moreover, is viewed as a catastrophic act of God or a sudden stroke of bad luck for which one cannot plan. The increased potential for damage resulting from shoreline engineering is also ignored in most cost evaluations. In fact, very few

Fig. 3.6. The failed seawall constructed on South Padre Island changed the equilibrium of the beach.

money—public funds—to protect the beach. Because these property owners do not constitute the general public, their personal interests do not warrant the large expenditures of public money required for shoreline stabilization.

Exceptions to this rule are the beaches near large metropolitan areas. The combination of extensive high-rise development and heavy beach use (100,000 or more people per day) affords ample

Fig. 3.7. Monmouth Beach, New Jersey, seawall.

shoreline-engineering projects would be funded at all if those controlling the purse strings realized that such "lines of defense" must be perpetual.

Once you begin shoreline engineering, you can't stop it! This statement, made by a city manager of a Long Island Sound, New York, community, is confirmed by shoreline history throughout the world. Because of the long-range damage caused to the beach it "protects," this engineering must be maintained indefinitely. Its failure to allow the sandy shoreline to migrate naturally results in a steepening of the beach profile, reduced sand supply, and therefore accelerated erosion (see chapter 2). Thus, once man has installed a shoreline structure, "better"—larger and more expensive—structures must subsequently be installed, only to suffer the same fate as their predecessors (fig. 3.7).

History shows us that there are two situations that may terminate shoreline engineering. First, a civilization may fail and no longer build and repair its structures. This was the case with the Romans, who built mighty seawalls. Second, a large storm may destroy a shoreline-stabilization system so thoroughly that people decide to stop trying. In America, however, such a storm is usually regarded as an engineering challenge and thus results in continued shoreline-stabilization projects. As noted earlier, rubble from two or more generations of seawalls remains off some New Jersey beaches!

The solutions

1. Design to live with the changing coastal environment. Don't fight nature with a "line of defense."
2. Consider all man-made structures near the shoreline *temporary*.
3. Accept as a last resort any engineering scheme for beach restoration or preservation, and then, only for metropolitan areas.
4. Base decisions affecting coastal development on the welfare of the public rather than the minority of shorefront property owners.
5. Let the lighthouse, beach cottage, motel, or hot dog stand fall when the time comes.

4. Selecting a site on a Texas beach

On September 8, 1900, a hurricane hit Galveston, which at that time was the fourth largest city in Texas, with a population of more than 37,000. The storm washed away two-thirds of the buildings in that city, at a cost that has never been fully calculated, and 6,000 people were killed. As many as 2,000 people may have been killed in other coastal areas.

It is safe to say that the Galveston disaster marked the beginning of America's concern with coastal hazards and how man can best coexist with the sea at the shoreline.

Have we learned our lesson? Could the 1900 hurricane repeat itself and cause great damage and loss of life once again? The answer is a clear *yes*. In fact, most coastal communities would be severely damaged by an extreme storm such as Hurricane Camille, which struck Mississippi in 1969. If Hurricane Allen (1980) had come ashore at its full force in a developed section of the Texas coast it would have equaled or exceeded the two-billion-dollar loss of Camille.

Considering the dangers, is coastal development a hopeless situation? Should we throw up our hands and walk away from the coast and allow any type of development to go on since it will likely be destroyed anyway? The answer is a clear *no*.

Over the last two decades we have learned a great deal about how beaches respond to storms and how islands evolve as sea level rises. Engineers have studied the response of various types of structures to storm conditions, and building codes have been devised to improve building safety. Most important of all, Texas geologists have spent years studying the Texas coast from Sabine Pass to the Rio Grande. As a result, the present condition of the Texas coast is no longer one of the mysteries of the sea. In fact, more information for public consumption has been published about the Texas coastal environment than for any other American coastal state. Furthermore, hurricanes are no longer a mystery. Gone are the days when a citizen of Galveston will first know about a hurricane when the winds reach 75 mph.

We know that there are ways to enhance the safety of coastal developments and coastal dwellers. We also know how to develop beach areas so that the beach is preserved for future generations. We have seen the mistakes of New Jersey and south Florida and we have learned from those mistakes.

If a hurricane such as Camille were to strike any Texas community head on it would certainly cause great destruction. But the chances of such a strong hurricane hitting any portion of our coast are slim and we can successfully prepare for lesser storms. Storms, however, aren't the only important problem affecting the Texas coastal resident. The worldwide sea-level rise is causing erosion along the Texas shore at rates typically ranging from 5 to 20 feet per year.

Because of widespread beach erosion, wave action, and storm

flooding, no dwelling site near a Texas beach is without risk. But intelligent and careful planning can result in selecting the sites of lowest risk, thus making the coast a safer place to live. That's what this chapter is all about. If man chooses to confront nature at the shore, we hope this chapter will at least make it clear what the consequences may be.

Nature's clues to danger at the beach

For the wary coastal dweller, nature holds many clues that can reveal much about the safety of a particular lot near the beach. Although it helps to be an expert on coastal processes it's not at all necessary. In fact, as one reads through the list below it is apparent that many of the indicators of site safety are simple common sense.

There are some aspects of site choice that common sense alone won't really solve. For example, if one is examining a site on the bay side of an island or peninsula, one almost always examines the site on a bright, sunny summer day. The bay is calm and waves small or nonexistent. It is difficult in the extreme for most people to imagine what the same bay looks like (and what the waves pounding the shore are like) during an intense winter storm.

Another aspect of site choice where common sense often fails is the long-range view of man's impact. For example, a long seawall in front of large buildings (such as on the North or South Padre beaches) may seem to coexist in perfect harmony with a broad beach; 20 to 60 years from now the beach in front of the wall almost surely will be gone as at Galveston and the wall probably will have been replaced by a much larger structure.

Although the erosion rates cited in this chapter are valid and well documented, many people are skeptical of them. Too often people fail to understand the significance of such erosion rates in terms of where the shoreline will be when their children or grandchildren inherit the house.

The wise landowner knows that more than natural forces are at work. The politics of a community play a major role in determining how the community will interact with nature. Many American coastal communities with summertime populations in the tens of thousands are controlled politically by a few hundred year-round residents. Therefore, it's important to understand the politics of a beach community.

Apart from the political climate, the most important natural clues to site selection on Texas barriers include elevation, vegetation, dune topography, soil types, and the general *terrain* (or coastal environment).

Elevation

Low, flat areas are flooded when water rises during a storm. These areas are also subject to waves washing over either from the bay side or the Gulf side of an island, especially the latter. Would common sense say otherwise? High areas are less prone to flooding and wave attack. So buy high, if possible!

Increasing elevation by trucking, bulldozing, or pumping sand is a common practice in Texas. Artificially elevated land is better than lower-elevation construction but is a poor substitute for naturally high land. If your site was elevated at the expense of beach and dune sand (a common practice) you may be in for trouble.

"High" elevations on the Texas coast are of the order of 15 to 25 feet and are virtually always atop dunes. You can purchase a U.S. Geological Survey map of the island (see appendix B) to find out the natural elevation of your site, or less preferably, you can estimate it by eye.

Vegetation

The presence or absence of vegetation is critical for sites in sand dunes. Lack of dune vegetation because of climate (South Padre) or even because of hiking paths and dune buggies may cause the dunes to begin shifting about. Small areas of active dunes can be stabilized, but large active dune fields such as on North and South Padre Island should be avoided (fig. 4.1).

Dunes

Nearly everywhere on the U.S. Atlantic and Gulf barrier coasts, dunes are the most important element in site safety. Sand dunes are usually responsible for site elevation and also for direct protection from storm waves and flooding. From a geological perspective, waves cannot attack and destroy a house until they have eroded and destroyed the dunes. In spite of this, some beach cottage owners have removed frontal dunes to improve the view of the sea. For example, dwellers of Gilchrist and Crystal Beach have removed some of their dunes, and these communities only

Fig. 4.1. Dune migration on North Padre Island.

had very small dunes to begin with. A walk up and down the beach near your proposed homesite may reveal whether dunes have been removed. If the character of the dune line changes in front of developments, sand may have been removed (fig. 4.2).

Soils

A look at the soil types may help you understand your choice of a homesite. If you have fresh shells mixed in with the sand, it may be overwash material brought in by storm waves. Washover areas are not a desirable homesite and at the very least call for construction on stilts. Shells are also present when material has been dredged up from the bay side to increase the elevation of a site. Because sand dunes are constructed by the forces of the wind, they rarely contain seashells of any size. Shells from the beach commonly have some brown coloration whereas shells pumped up from the lagoon tend to be black and white in color. Don't depend on this distinction entirely however.

Terrain

This is the broadest category of natural clues that the Texas coastal dweller must consider in deliberating over the safety of a homesite or proposed homesite. Terrain (or coastal environment) as we use it here refers to the shape of the land surface. The shape of the surface affords important clues to what processes affect the area. The wise property buyer will put on hiking boots and tramp around for a while and try to understand his terrain. By identifying and understanding the various environments of an island or coastal area one can determine the safety of a particular area for development.

Washover channels. These permanent features (figs. 2.4 and 2.7) are unique to the barrier islands of Texas. Other American barrier islands do not have them and geologists are not certain why this is the case. Washover channels mark the route by which water flows all the way across the island during a storm (fig. 4.3).

Fig. 4.2. Removal of sand dunes in developed areas caused greater penetration of overwash from Hurricane Carla (1961) on Bolivar Peninsula. Source: National Oceanic and Atmospheric Administration.

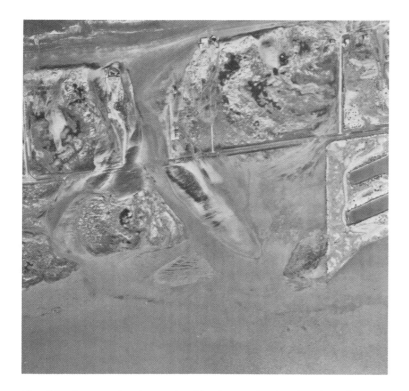

Fig. 4.3. Overwash of road by Hurricane Beulah in 1967. Source: Texas Department of Highways and Transportation.

These channels are occupied repeatedly in storm after storm. Hence, construction should never be carried out in such a channel without some modification of the surface elevation.

It is sometimes not easy to recognize washover channels. Near the Gulf side of the island, they range from tens to hundreds of yards in width. The channels tend to flare or widen toward the bay or lagoon side of the island. Frequently on the bay side, the washover channels are marked by long, narrow, standing bodies of water. Be careful; all long, narrow, standing bodies of water are not connected to washover channels.

Perhaps the best way to identify washover channels is to stand on the upper beach and look toward the bay. If you can see much of the back side of the island and the bay, you are probably looking down the throat of a washover channel. This may not work, however, on the southern portions of Padre Island; between storms small unvegetated (unstabilized) sand dunes creep into the throat of overwash channels making them difficult to spot from the beach. Such sand dunes afford little protection during a storm and are quickly washed away by the waves and flowing waters.

An example of construction in an overwash channel can be found on South Padre Island. Here a condominium is virtually centered in a washover channel. The condominium builders increased the elevation of the site with fill material which will prevent flooding during small washover events. The fill will not prevent large washovers, however, and when they do occur the condominium will likely cause diversion of flood water into adjacent developments. Nature has presented Texas barrier island dwellers with a serious but easily identifiable hazard to development. On barrier islands outside of Texas, the location of storm overwash is much more difficult to predict.

Backbarrier flats. Texas barrier islands usually have broad, flat areas of low elevation on the bayshore side referred to as *backbarrier* or *vegetated barrier flats* (fig. 2.4). The principal hazards facing homeowners here are wave action and flooding from the bay side. Don't be fooled by the typical calm appearance of the bay. Under the right wind conditions large waves capable of causing destruction can be generated. The baysides of most Texas barrier islands are eroding, but usually bulkheads are constructed adjacent to developed areas to retard erosion. Bulkheads can successfully retard erosion caused by the action of waves in bays, but don't expect the same success with bulkheads on the Gulf side of the island.

The main rule for living on the backbarrier flat of a Texas island is to elevate the house. This is frequently done by using dredged material from adjacent boat canals to raise the ground elevation or by building on stilts; sometimes both means are used.

Primary dunes. Dunes in the ridge closest to the beach are called *primary dunes* or *fore-island dunes* (fig. 2.4). Primary dunes are the natural main line of defense against erosion and storm damage to man-made structures. Ideally, all construction should be landward of the primary dunes and construction should never involve removal of sand from the primary-dune system.

Sand in the primary dunes came originally from the beach. The exchange of sand between beach and dune is a continuous process, and construction of seawalls on the beach prevents addition of sand to the dunes from the beach. The primary dunes also serve as a reservoir of sand to be used by the beach during major storms (see chapter 2).

Vegetation stabilizes (holds in place) the primary dunes. Dune buggies, foot traffic, drought, and fire destroy dune vegetation and may destabilize the dunes. Needless to say the primary dunes are the most susceptible of all island dunes to destabilization, that is, loss of vegetation causing the dunes to move again.

If dunes are destroyed or threatened, there are some remedies that can be taken to restore them artificially. Planting dune grass or sea oats in bare areas serves to stabilize existing dunes and encourages additional dune growth. Sand fencing (snow fencing) is commonly used to trap sand and to increase the size of the primary dunes. The success of sand fencing depends largely on whether sand is presently being blown inland from the beach.

Marshes. Marshes (fig. 2.4) are prolific breeding areas for many marine organisms including shrimp and fish. Their extensive shallows provide considerable protection against wave erosion of the bay shoreline. In the past, marshes have been filled to expand land areas on which to build. (Examples are found on Bolivar Peninsula and Galveston Island.) Buried marshes provide poor support for buildings, and septic systems usually do not function properly at such sites. Hence, construction in marshes should be avoided. Also it is illegal to dredge, drain, or fill a marsh without a permit.

Inlets (passes). Inlets are channels that separate islands (fig. 1.1). They are the means by which water is exchanged between the open ocean and bays. They are also the means by which ships enter and leave the ports of Texas.

Inlets, commonly called passes in Texas, tend to be dynamic. Under natural conditions they sometimes widen and narrow, and at other times they migrate up or down the coast. Aransas Pass was migrating to the south at 250 feet per year before jetties were constructed. Most of the inlets in Texas have been stabilized by jetties, the long rock walls extending seaward, sometimes for miles. There are two artificial inlets in Texas constructed for boating or shipping purposes. One is Mansfield Channel near South Padre and the other is Matagorda Ship Channel on Matagorda Peninsula (fig. 1.1). Two other examples of artificial inlets are also unique to Texas: Rollover Fish Pass on Bolivar Peninsula and the fish pass on Mustang Island. It is not clear whether such passes actually have significantly enhanced fishing, and the passes have proved to be costly to maintain.

If you intend to live on an island with unstabilized inlets (an inlet or pass without jetties), a good rule is to stay 0.5 mile or more away from the inlet. If the inlet has been jettied, a good general rule is don't buy near the beach. Changes in shoreline position can be rapid and extreme near some jetties.

The problems people cause at the beach

Nature creates a lot of headaches for man at the shore. Inevitably man exacerbates the natural hazards by choosing to confront nature. Thus we not only increase the hazards from natural forces, we also create our own additional hazards.

The major people-created problems on the Texas coast can be grouped under the following headings: stabilization of shorelines, construction, water and sewage, finger canals, and escape routes.

Stabilization of shorelines

Shoreline-stabilization projects are discussed in chapter 3 and need not be discussed again here. In short, attempts to halt shoreline erosion ultimately result in destruction of the beach at great cost to taxpayers and individual homeowners. Furthermore, such attempts at stabilization, unless they are of the magnitude of the Galveston seawall, are usually damaged or destroyed by the sea, requiring additional and continuous expenditures for maintaining and rebuilding.

Construction

Tips on construction techniques are discussed in chapter 6. The main point here is that unless your house is well sited and constructed it could be lost in the next storm. There is no way to construct a house to guarantee its survival in a hurricane such as Camille. Although such extreme storms are the exception, it is worth the slight added construction cost to increase the likelihood of house survival in the more frequent smaller hurricanes. Perhaps an equally important consideration is the construction of the house of your uninformed or miserly neighbor. His house may fall apart in a storm and batter yours down.

Hurricane Camille illustrates well the wisdom of quality construction. The Mississippi shore home of Jefferson Davis, President of the Confederacy, survived this storm with minor damage while many other houses in the same community disappeared altogether.

Water and sewage

Most shore communities of Texas are served by central water-supply systems and sewage disposal plants. Homes in isolated areas away from island communities, however, often have septic tank systems for sewage disposal.

Apparently some older homes in towns now served by sewage systems still use septic tanks, and a few communities such as Sargent Beach and Surfside and homes on Bolivar Peninsula use septic tank systems entirely. Such systems consist of a holding tank in which solids settle and sewage is biologically broken down. Overflow from the tank flows into a drain field that allows the water to percolate into the soil where the water is filtered and purified naturally. Septic tanks are perfectly suitable means of sewage disposal providing the development doesn't become so dense that local soils become saturated with septic tank effluent. At present, this does not appear to be a problem in Texas beach communities. Along the Atlantic coast, however, septic tanks are blamed for the closing of thousands of acres of oyster and clam fishing grounds.

Perhaps the most important sites of potential pollution on the Texas coast are the long finger-canal developments in communities that use septic tanks for sewage disposal (e.g., Bolivar Peninsula).

Finger canals

Finger canal is the term applied to a waterway or channel dug from the lagoon or bay side of an island into the island proper for the purpose of providing residents with a waterfront lot. Canals can be made by excavation alone or by a combination of excavation and infill of adjacent low-lying areas (usually marshes). Finger canals can be found in most Texas coastal areas including Bolivar Peninsula, Galveston Island, Mustang Island, and Padre Island.

The major problems associated with finger canals are the (1) lowering of the groundwater table; (2) pollution of groundwater by seepage of salt or brackish canal water into the groundwater table; (3) pollution of canal water by septic seepage into the canal; (4) pollution of canal water by stagnation due to lack of tidal flushing or poor circulation with bay waters; (5) fish kills generated by higher canal-water temperatures; and (6) fish kills generated by nutrient overloading (algal blooms) and deoxygenation of water (fig. 4.4).

Bad odors, flotsam of dead fish and algal scum, and contamination of adjacent shellfishing grounds are symptomatic of polluted

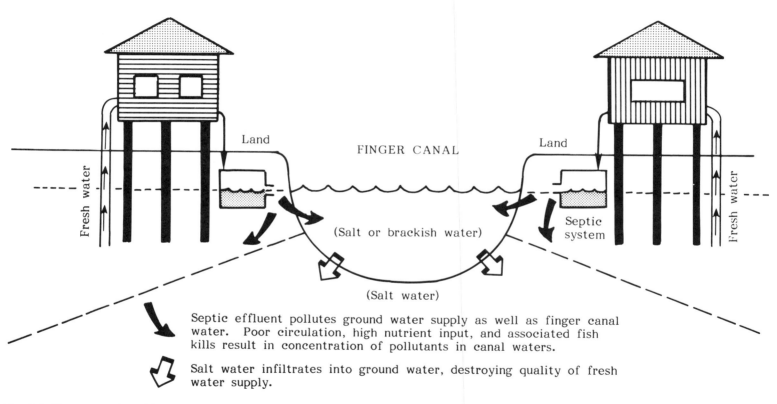

Fig. 4.4. Finger-canal problems.

canal water. Thus, finger canals often become health hazards or simply places where it is unpleasant to live. Residents along some older Florida finger canals have built walls to separate their cottages from the canal!

Texas finger canals have not been reported to be in polluted condition as yet. However, the Florida experience indicates that the problems usually don't appear until a few years after complete development has occurred. Short canals are generally much less likely to become polluted than long ones. Tributary canals are considerably more likely to experience pollution than main canals.

In some instances finger canals cross most of the island and may offer a path of least resistance to storm waves; they are therefore potential locations for new inlets or at least concentration of eroding floodwaters. An example of this type of problem can be seen on South Padre (fig. 4.3). On North Padre, finger canals extend from a major washover channel (Packery Channel) that was formerly a tidal inlet. Basically this is a way of bringing the washover right to one's doorstep in the next storm.

Escape routes

Because of the threat of hurricanes, an escape route must exist that will permit people to leave an island and get safely inland within a reasonable length of time. The presence of a ready escape route near a building site is essential to site safety, especially in high-rise or other high-density areas where the number of people to be evacuated, transported, and housed elsewhere is greatly increased.

Select an escape route ahead of time. Check to see if any part of it crosses a low elevation or is subject to blockage by overwash or flooding; if so, seek an alternate route. Note whether there are bridges along the route. Remember that some residents will be evacuating pleasure boats and that fishing boats will be seeking safer waters; thus, drawbridges will be accommodating both boats and automobiles.

Re-evaluate the escape route you have chosen periodically—especially if the area in which you live has grown. With more people using the route, it may not be as satisfactory as you once thought it was.

Use the escape route early. Be aware that most Texas islands have only one route for escape to the mainland. In the event of a hurricane warning, leave the island immediately; do not wait until the route is blocked or flooded. Anyone who has experienced the evacuation of a community knows of the chaos at such bottlenecks. Depend on it; excited drivers will cause wrecks, run out of gas, have flat tires, and cars of frightened occupants will be lined up for miles behind them. Be sure to plan where you will go. Keep alternative destinations in mind in case you find the original refuge filled or in danger.

Consider the problems caused by Hurricane Carmen, which hit the Gulf Coast in September 1974. Over 75,000 people are said to have evacuated from what were thought to be the danger areas in Louisiana and Mississippi. The traffic was bumper to bumper on the few roads leading north. One accident backed up traffic for 19 miles. Motel lobbies were filled with people looking for a place to stay; all rooms were taken. Weary people were forced to continue traveling north until they found available space. Similar experiences should be anticipated for Texas.

The National Oceanographic and Atmospheric Administration has published evacuation maps for the coast of Texas (see appendix B). The maps give the elevation of low points in evacuation routes from barrier islands. The low points are the most dangerous areas you must watch out for as you are fleeing an approaching hurricane.

Storm evacuation for some important developing areas along the Texas coast can be summarized as follows. Escape from Bolivar Peninsula is via a road of very low elevation that is frequently flooded. Escape routes from Galveston are generally good by coastal standards **but** a tremendous number of people may have to escape simultaneously. Mustang and North Padre Island residents must escape via the Kennedy Causeway or the Port Aransas Ferry. The elevation over much of the causeway is very low; it has the lowest elevation of any escape route for a coastal community of significant size in Texas. For South Padre the escape route is the Queen Isabella Causeway (fig. 4.5), a high bridge; however, the island road leading to the bridge is low and may be flooded hours before the storm nears the coast.

Conclusion: **leave early**.

Fig. 4.5. Queen Isabella Causeway, South Padre Island. Note low elevation of approaches. Source: Texas Coastal and Marine Council.

The site: checklist for safety evaluation

The following is a guide to evaluating your potential homesite. Find your shoreline area of interest later in this chapter, and using the general information given in chapters 1, 2, and 3, plus the checklist given below, evaluate your site.

1. Site elevation is above anticipated storm-surge level.
2. Site is behind a natural protective barrier such as a ridge of sand dunes.
3. Site is in an area of shoreline growth (accretion) or low

shoreline erosion. (Evidence of an eroding shoreline includes: a. sand bluff or dune scarp—small cliff—at back of beach, b. mud exposed on beach, c. threatened man-made structures, and d. protective devices such as seawalls, groins, or artificially emplaced sand.)
4. Site is located on a portion of the island backed by a barrier flat or salt marsh.
5. Site has good access to an evacuation route, which is adequate to handle mass evacuation under storm conditions.
6. Site is away from low, narrow portions of the island.
7. Site is not in an area of historic overwash.
8. Site is well away from a migrating inlet.
9. Site is in a vegetated area that suggests stability.
10. Site drains water readily (even in wet season).
11. Water supply is adequate and uncontaminated. Water and sewage systems are adequate for present demands and anticipated growth.
12. Soil and elevation are suitable for efficient septic-tank operation.
13. No compactable layers such as mud are present in soil below footings. (Site is not on a buried salt marsh.)
14. Adjacent buildings are adequately spaced and of sound construction.
15. Federal flood insurance is available.
16. Building codes exist and are really enforced.
17. The year-round residents (who are the electorate) agree with your outlook on the future of the community.

Site analysis: the coast of Texas

Sabine Pass to the vicinity of Sea Rim State Park (figures 4.6 and 4.7)

This segment of mainland shoreline is characterized by low and broad mud flats covered by marsh vegetation with intervening grass-covered cheniers or beach ridges. The ridges are composed of sand and are slightly higher in elevation than the surrounding marsh. If one judges from the curvature of these topographic features, he will conclude that Sabine Pass has remained in essentially the same position for the past few thousand years. The inlet has remained stable largely because of the mud in the channel banks. Today the channel is periodically dredged and maintained for deep-draft vessels and the channel has been extended onto the inner shelf between long, rock jetties.

This area is undergoing extreme shoreline erosion, especially near Sabine Pass. The beach is very narrow and consists of a thin layer of shell and sand overlying eroding marsh mud. The mud below the sand makes driving on this beach very treacherous. The shoreline is eroding at the rate of 10 to 30 feet per year, and possibly the jetties at the mouth of Sabine Pass play a role in this high rate by not allowing the longshore currents to carry in sediments. However, the jetties acted as sediment traps and caused rapid advancement of the adjacent shorelines after they were initially constructed in 1885.

There are no dunes along this shoreline and elevations are

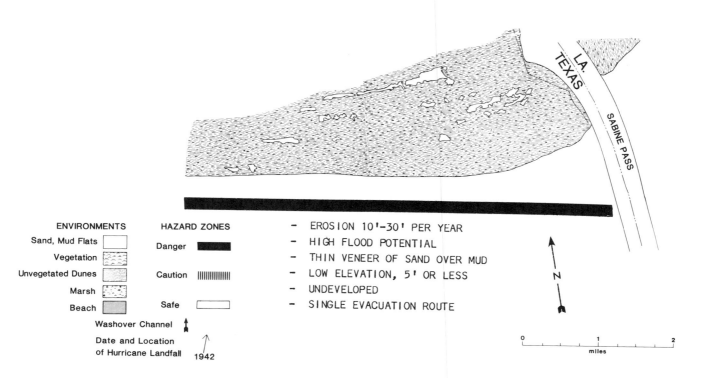

Fig. 4.6. Site analysis: Gulf beach near Sabine Pass.

52 Living with the Texas shore

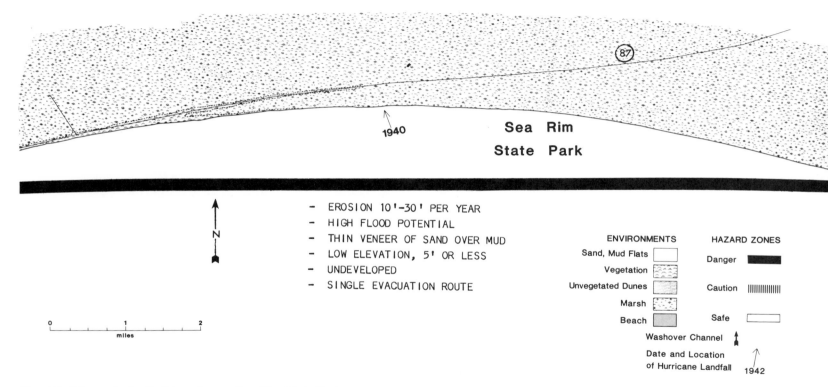

Fig. 4.7. Site analysis: Gulf beach near Sea Rim State Park.

extremely low. Consequently, the entire area including Highway 87 is overwashed even during minor storms. Because the road is easily flooded there is no safe escape route from the area. The most recent severe flooding occurred during Hurricane Carla (1961) when flood waters extended 10 to 12 miles inland from the shoreline. The area is generally unsuited for development owing to the high hazard potential. Moreover, preservation of marsh areas and poor beach conditions have prevented development of the area except for facilities that support onshore and offshore drilling operations.

Sea Rim State Park vicinity to High Island vicinity (figures 4.7 and 4.8)

The beach along this shoreline segment is slightly broader than the beach near Sabine Pass but it is still basically a thin veneer of sand over mud. Much of the beach is covered by coarse material composed mainly of rock fragments and old oyster and clam shells that were eroded from the adjacent offshore sediments and tossed up on the beach. At Sea Rim State Park grass plantings have aided in constructing a foredune ridge. West of Sea Rim State Park the dunes are low and discontinuous. In some areas the low dunes that separate the beach from the road were artificially constructed from overwashed sand scraped off the roads following storms.

The beach and adjacent marshes have elevations of 5 feet or less and are easily flooded. In fact, storm waters from Hurricane Carla (1961) extended inland more than 10 miles. Some low-lying areas are inundated frequently even by minor storms. For example, Highway 87 (fig. 4.8), the only evacuation route east of High Island, is overwashed repeatedly and closed periodically because of high tides associated with the passage of northers or winter storms. This same road is also vulnerable to shoreline erosion which has consistently averaged between 5 and 10 feet per year for the past 100 years (fig. 4.9). Today this coastal highway is at the water's edge in some areas and needs to be moved inland just as it was in the 1930s when relocation was required because of beach erosion.

Although the marsh is not right at the beach along this shoreline, the vegetated area with slightly higher elevations that separates the marsh from the beach is very narrow. The high rate of erosion, narrow beach width (50 to 75 feet), low elevation, and resulting widespread overwash make this area unsuitable for development, and except for oil-field facilities the area is undeveloped.

High Island vicinity to Rollover Pass (figures 4.8 and 4.10)

This short beach segment (9.5 miles) differs from the preceding shoreline stretch only in its most recent erosion rate and locally higher elevations of High Island which is the surface expression of a salt dome.

Development is concentrated at the extreme ends of this beach segment and is of two different types. At High Island, surface equipment and buildings associated with hydrocarbon production are prominent but are limited to the area of oil- and gas-field operations. In contrast, the development near Rollover Pass con-

54 Living with the Texas shore

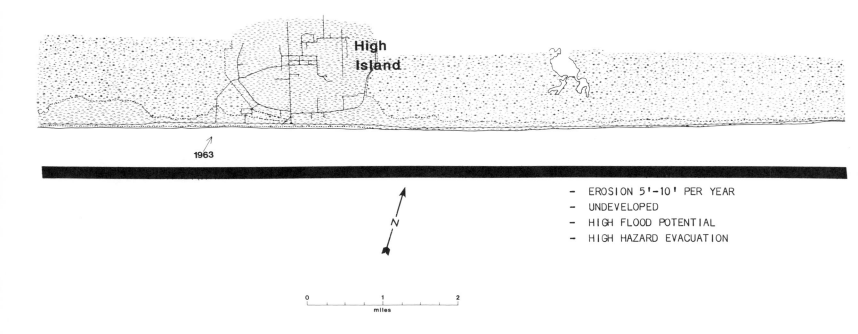

Fig. 4.8. Site analysis: Sea Rim State Park to High Island.

4. Selecting a site   55

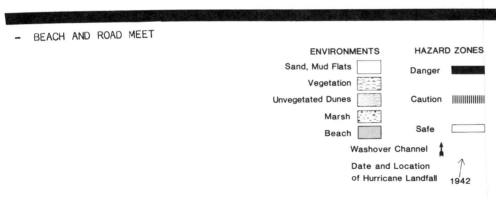

- BEACH AND ROAD MEET

ENVIRONMENTS   HAZARD ZONES
Sand, Mud Flats □   Danger ■
Vegetation   Caution ||||||||||
Unvegetated Dunes
Marsh   Safe □
Beach
Washover Channel ↑
Date and Location
of Hurricane Landfall ↑1942

- EROSION 5'-10' PER YEAR
- LOW ELEVATION, 5' OR LESS
- LOW, DISCONTINUOUS DUNES
- SAND BEACH
- HIGH FLOOD POTENTIAL
- OILFIELD DEVELOPMENT ONLY

Fig. 4.9. Road destroyed by beach erosion near High Island.

sists of wooden houses on stilts mostly located along the beach or adjacent to extremely long finger canals that were dredged through highly productive marshes. These narrow canals generally have poor circulation and have potential for pollution from septic tank effluent when development becomes dense. The poor circulation in the finger canals is a result of the length and orientation of the canals, the restrictions to tidal exchange presented by the adjoining Intracoastal Waterway, and the low tide range of about 1.5 feet. Only wind-generated currents and storm tides cause flushing of these canals. The entire area is less than 5 feet in elevation, so there is considerable danger from flooding. During Hurricane Carla (1961) the town of Gilchrist was completely inundated and washed over. Overall erosion rates are 5 to 10 feet per year. The wooden stilts on which the homes are built provide protection from flooding but not erosion. Abandoned cottages on the beach with exposed septic tanks attest to the long-term shoreline retreat in the area.

In 1955, Rollover Pass was cut to provide a shorter route for fish migration between East Bay and the Gulf. Shortly after the pass was opened, channel and bank erosion were so severe that the Corps of Engineers closed the inlet temporarily until steel bulkheads could be constructed. The pass is probably not causing undue erosion problems except immediately adjacent to the inlet. One aspect to consider, however, is that sand flows through the inlet and is deposited in Rollover Bay. This sand would normally have gone to the beach westward along Bolivar Peninsula where it would have reduced erosion or contributed to accretion near the Bolivar Roads Jetty. Moreover, the pass could become the focus of large waves because of greater depths in the channel compared to adjacent land areas.

Rollover Pass to Crystal Beach vicinity (figures 4.10 and 4.11)

This shoreline segment marks the beginning of Bolivar Peninsula—a true barrier island in the sense that it is surrounded by water. Accessibility to the adjacent East Bay and Intracoastal Waterway encourages finger canal development, in contrast to the mainland area between High Island and Sabine Pass. Because of poor circulation, finger canals may eventually be polluted from septic tanks when development is heavy. Canals that connect with East Bay have the lowest potential for pollution, whereas small tributary canals off main canals have the greatest potential for pollution. Long-term (100-year) erosion rates for the area have averaged between 0 and 5 feet per year with recent rates up to 10 feet per year. Because of the relentless shoreline retreat, gas wells drilled in the 1950s at ground level and behind the dunes are now exposed on the beach and in the surf zone. Steel and wooden cages were constructed to protect the wellheads from cars driving on the beach and from battering by floating debris (fig. 4.12).

Although shoreline erosion persists from year to year, dramatic beach losses associated with storms are most notable. The Gulf shoreline near Rollover Pass eroded 50 to 60 feet during Hurricane Audrey (1957) and an equal amount during Hurricane Carla (1961). Even tropical storm Delia (1973) removed approximately 60 cubic feet of sand per linear foot of beach in the vicinity of Rollover Pass.

Development here is mainly single-family recreational homes built mostly on land with less than 5 feet of elevation. For comparison, Hurricane Carla caused flooding of 9 feet. A few narrow grass-covered beach ridges are 10 feet or slightly higher in elevation. Flood potential during hurricanes is high but washover potential is less than in areas to the northeast. In general, storm washover is restricted to a band that is parallel to the beach and about 500 feet wide. In developed areas washover may penetrate much farther due to destruction of the dunes. As shown in figure 4.2, the removal of dunes was clearly responsible for greater washover penetration by Hurricane Carla. Evacuation from this area is precarious considering that the single overland route is easily flooded near Rollover Pass and the alternate route involves the ferry at Bolivar Roads which may be undependable because service can be disrupted by high waves.

Crystal Beach vicinity to Bolivar Roads (figure 4.11)

This barrier segment is perhaps the safest area for development on Bolivar Peninsula. The beach here is wide and growing seaward, so erosion is not a problem. This accretion is partly natural and partly the result of human activities. The broad western half of Bolivar Peninsula is marked by grass-covered beach ridges and intervening swales (depressions) filled with fresh-water marsh. The resulting ridge and swale topography shows that the barrier has built seaward during the past few thousand years.

The most recent shoreline accretion on the Peninsula's western tip can be attributed to sediment supplied by beach erosion and trapped by the north jetty at Bolivar Roads. Over 28 million cubic yards of sand have been added to the beach and along the jetties by coastal processes since jetty construction in 1876. The Galveston jetties provide some shore-front protection by reducing the height of waves generated by minor storms centered to the southwest. Potential for flooding is high; most development is on land that is

58 Living with the Texas shore

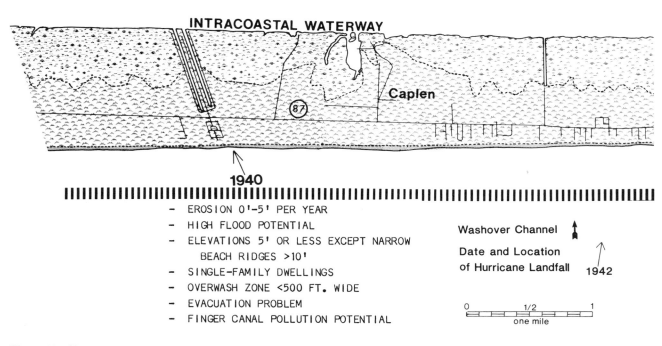

Fig. 4.10. Site analysis: Bolivar Peninsula from Gilchrist to Caplen.

4. Selecting a site   59

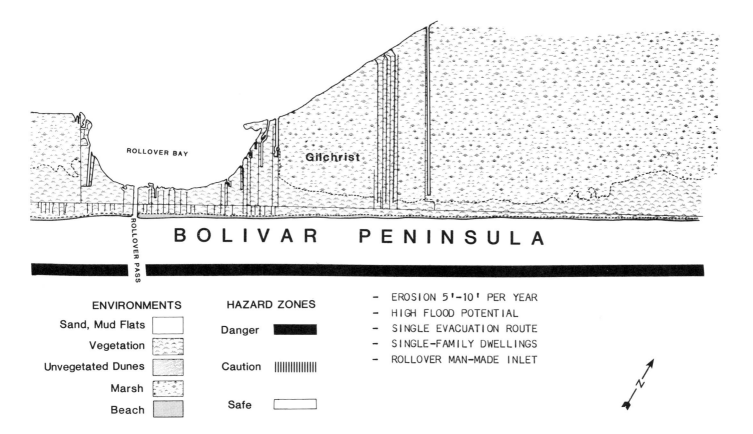

60 Living with the Texas shore

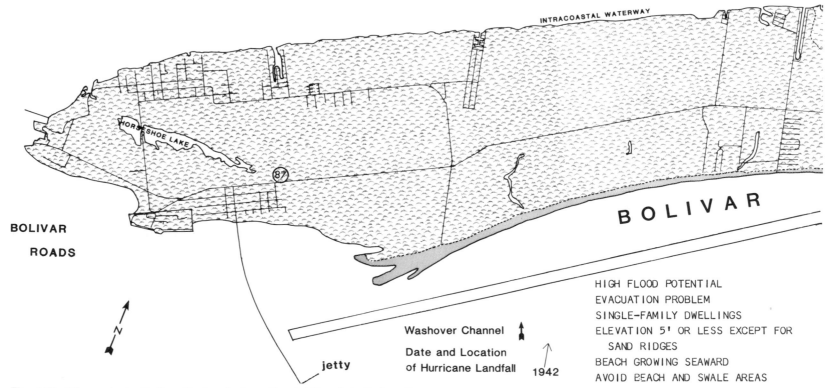

Fig. 4.11. Site analysis: Bolivar Peninsula from Crystal Beach to Bolivar Roads.

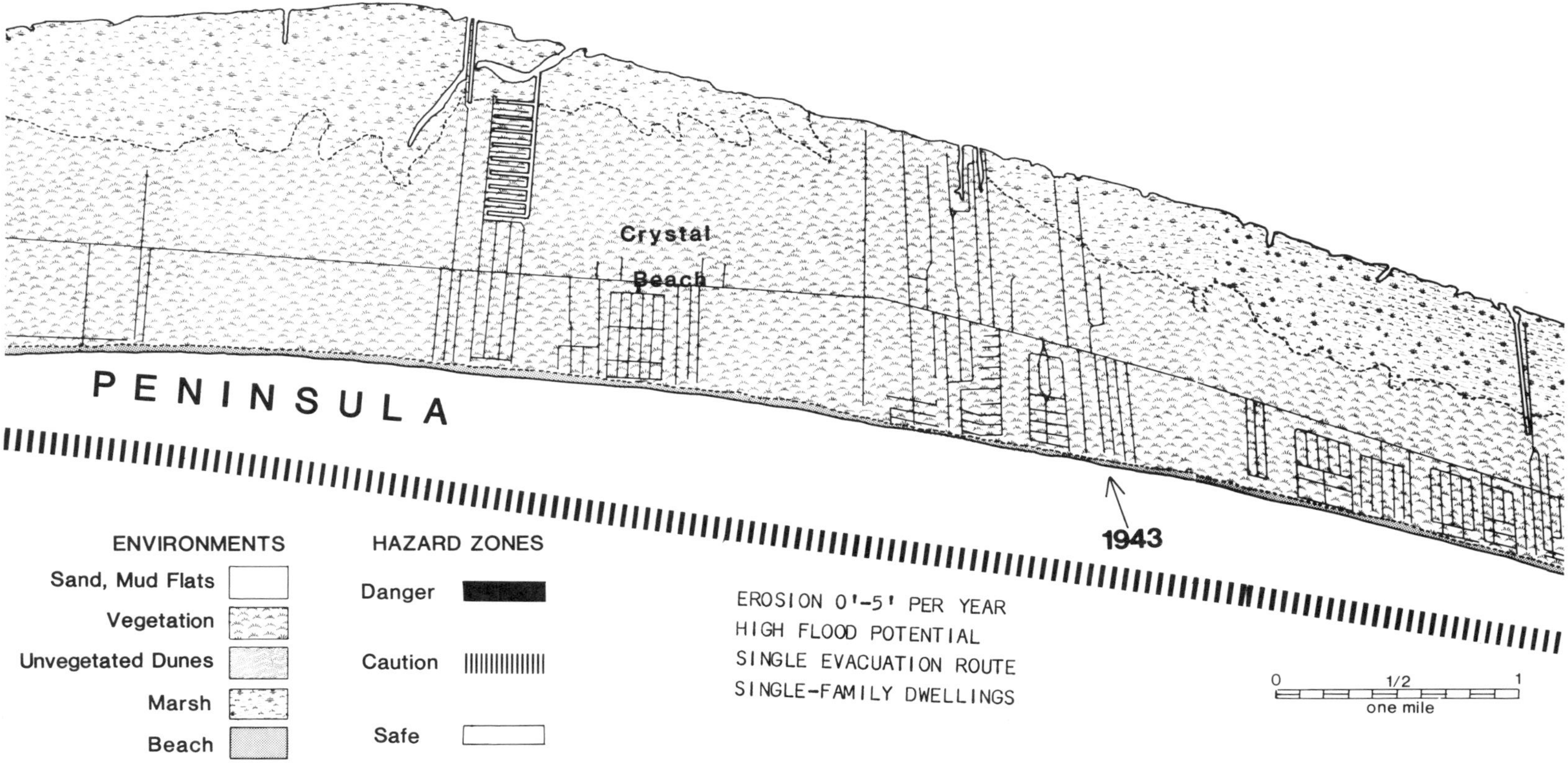

4. Selecting a site 61

PENINSULA

Crystal Beach

1943

ENVIRONMENTS
- Sand, Mud Flats
- Vegetation
- Unvegetated Dunes
- Marsh
- Beach

HAZARD ZONES
- Danger
- Caution
- Safe

EROSION 0'-5' PER YEAR
HIGH FLOOD POTENTIAL
SINGLE EVACUATION ROUTE
SINGLE-FAMILY DWELLINGS

0   1/2   1
one mile

Fig. 4.12. Wellhead in surf zone of Bolivar Peninsula.

5 feet or less in elevation. Because of these low elevations, the entire area was inundated by Hurricane Carla. The development in this area is mostly recreational homes built on stilts to accommodate the flooding. The relatively broad width of the peninsula reduces the danger of scour and overwash except near the beach. Residents along the beach could maintain the safest possible site by not removing the dunes.

In this century alone storms have damaged the upper Texas coast in 1900, 1915, 1918, 1934, 1938, 1940, 1941, 1942, 1943, 1946, 1947, 1955, 1957 (Audrey), 1959 (Debra), 1961 (Carla), 1963 (Cindy), 1970 (Felice), and 1973 (Delia). Much of Bolivar Peninsula has been developed since the most recent major storms hit the area, and many current residents may not be aware of the massive destruction that occurs within a few hours.

Evacuation from the area may be a problem. The primary evacuation route depends on the ferry that operates between Port Bolivar and Galveston. High waves can disrupt ferry service, thereby eliminating the primary evacuation route. The only alternative to the ferry is Highway 87 near High Island, and this route is not dependable because of the low elevation, and consequent high potential for flooding, of the highway.

Galveston Island (figures 4.13, 4.14, and 4.15)

Galveston Island is possibly one of the best known barrier islands in America because of its long history, attractiveness to tourists, and prominence in a popular song. It may also be the first Texas barrier to become completely developed. As a major port, recreational resort, and medical center, Galveston has reached a critical mass (population 75,000) and is experiencing considerable pressure to grow even more rapidly. Until the 1970s the island offered a variety of land uses including high-density commercial and residential development within the city, as well as cattle ranching and scattered communities of summer homes on the western two-thirds of the island. Today the entire island falls under the corporate jurisdiction of the city, and high-density development has spread far beyond the seawall.

Galveston was incorporated in 1838, making it the oldest city on a Texas barrier island. At the turn of the twentieth century Galveston was best known for the almost complete destruction of the city during the 1900 hurricane, a storm that killed over 6,000 people and caused an estimated $25 million in damage. For the number of deaths this was the worst weather-related disaster in our history. Thousands of inhabitants fled the island although evacuation was difficult. Storm evacuation routes today are substantially better but they are still hazardous. The residents and visitors have essentially three options. The primary evacuation route is Interstate 45 to Houston, but apart from the usual problems associated with evacuation (overcrowding, wrecks, stalled vehicles), the approaches to the bridge are low and could flood. High flood-potential of the evacuation route and bridge approaches is also a problem with the second option, which involves driving the length of Galveston Island to Freeport. Another alternative proposed in some coastal areas is called vertical evacuation; this is the temporary use of high, well-built structures for storm protection. The wise coastal dweller will closely monitor the weather advisories and leave early to avoid the rush.

In addition to the 1900 storm and 1915 hurricane, which caused additional damages estimated over $50 million, seven other storms have struck Galveston this century. The most recent storm that caused major damage to Galveston was Hurricane Carla in 1961.

Galveston jetty to Galveston seawall (figure 4.13)

The eastern tip of Galveston Island has been drastically altered by dredging and jetty construction designed to improve the harbor entrance. Accretion of East Beach attendant with jetty construction has also been substantial and has ranged from 200 to 7,000 feet in slightly less than 100 years. This rapid seaward growth produced a broad sand flat with an elevation of about 3 feet, comparable to most backbeach elevations. Low, hummocky (hilly) dunes formed after the present shoreline position was reached in the mid-1950s.

At present this beach area is generally stable and is afforded some protection by the Galveston jetties. Since 1970, however, there has been slight erosion on the order of 0 to 5 feet per year. Although there is some high-rise condominium construction, most of the East Beach area is used for recreational purposes including a public beach and a seasonally operated amusement park. The area, which is flood-prone because of its low elevations, was completely inundated by a 9-foot storm surge that accompanied Hurricane Carla in 1961.

A minor debate has ensued over the disposition of wind-blown sand that accumulates on the beach and interferes with vehicular traffic. The city has recommended removal of the sand from East Beach and nourishment of the pocket beaches in front of the seawall. On the other hand, state officials have recommended that the sand be used on East Beach to nourish the low dunes.

64 Living with the Texas shore

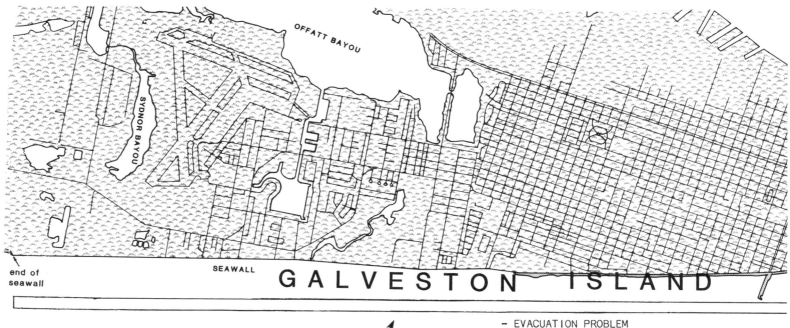

- EVACUATION PROBLEM
- NARROW RECREATION BEACH
- SEAWALL - GOOD OVERWASH PROTECTION FROM MOST STORMS
- HIGH DENSITY COMMERCIAL DEVELOPMENT
- NO SHORELINE EROSION
- HIGH FLOOD POTENTIAL

Fig. 4.13. Site analysis: Galveston Island from Bolivar Roads to end of the seawall.

4. Selecting a site 65

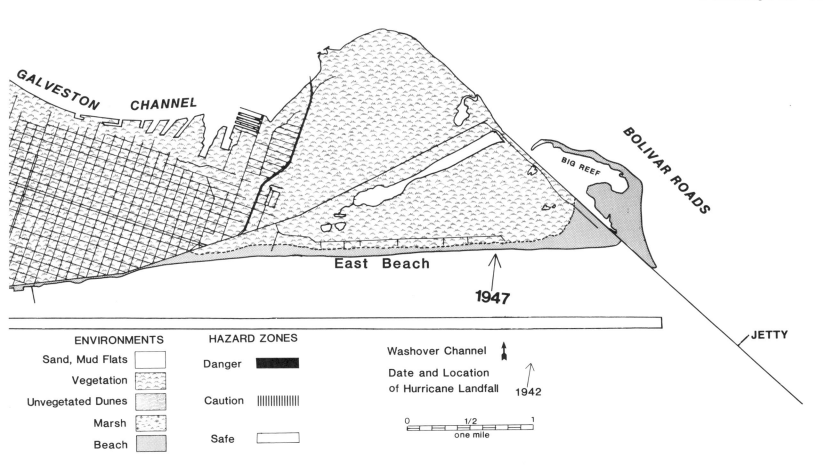

66 Living with the Texas shore

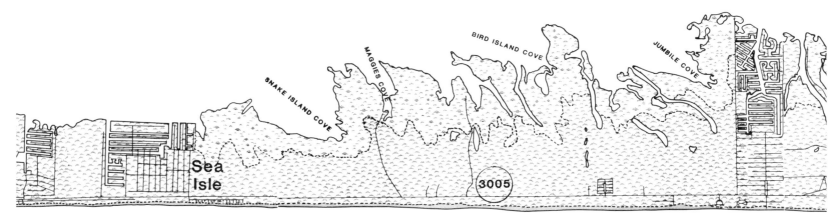

- EROSION 5'-10' PER YEAR
- ELEVATIONS ABOUT 5'
- LOW DISCONTINUOUS DUNES
- HIGH FLOOD POTENTIAL
- SINGLE EVACUATION ROUTE
- SINGLE AND MULTIFAMILY DWELLINGS
- NARROW ISLAND
- BAYSHORE EROSION

- EROSION 0'-5' PER YEAR
- LOW DISCONTINUOUS DUNES
- HIGH FLOOD POTENTIAL
- ELEVATIONS 5'-10'
- SINGLE-FAMILY DWELLINGS
- SINGLE EVACUATION ROUTE

Fig. 4.14. Site analysis: Galveston Island from end of seawall to Sea Island.

4. Selecting a site

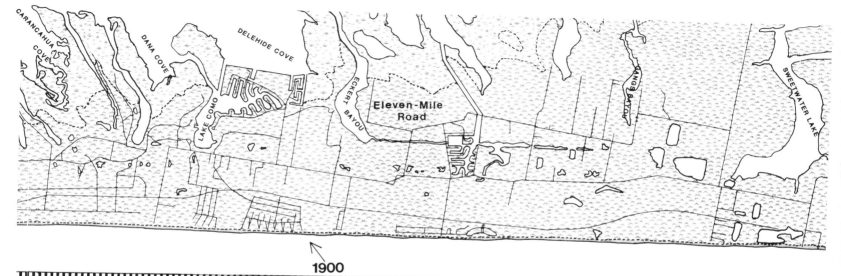

ENVIRONMENTS
- Sand, Mud Flats
- Vegetation
- Unvegetated Dunes
- Marsh
- Beach

HAZARD ZONES
- Danger
- Caution
- Safe

- EROSION 5'-10' PER YEAR
- HIGH FLOOD POTENTIAL
- ELEVATIONS 5'-10'
- SINGLE-FAMILY DWELLINGS
- LOW DISCONTINUOUS DUNES
- RAPID EROSION POSSIBLY DUE TO ADJACENT SEAWALL
- EVACUATION PROBLEM
- AVOID BEACH AND SWALE AREAS

Washover Channel
Date and Location of Hurricane Landfall 1942

0   1/2   1
one mile

68 Living with the Texas shore

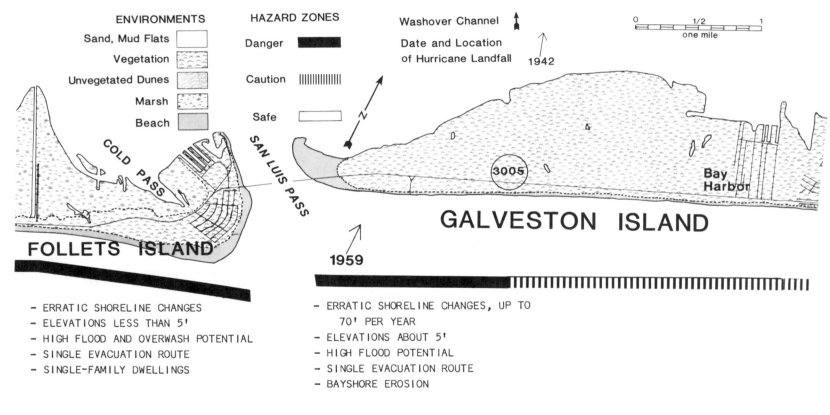

Fig. 4.15. Site analysis: western end of Galveston Island and eastern end of Follets Island.

4. Selecting a site 69

Fig. 4.16. Retreating beach, west end of Galveston seawall.

Galveston seawall (figure 4.13)

The 17-foot-high Galveston seawall is perhaps the best built seawall on any barrier island in the United States. The first segment of the seawall cost six million dollars to construct and an additional six million dollars was spent to increase the city's grade and elevation. The wall was built in 1902 in response to the destruction of Galveston by the 1900 hurricane. At that time a wide beach was present. Even as late as 1965, cars were driven in front of the western end of the seawall. Since then, however, the beach has largely disappeared (fig. 3.5). Recent studies by the Corps of Engineers indicate that the offshore sediment is mostly mud and offers very little sand for beach replenishment. Maximum beach erosion in front of the seawall occurs at the western end where 1,000 feet of beach-dunes once existed (fig. 4.16). The western end of the seawall was severely damaged by Hurricane Carla but overall it has withstood the test of time and has successfully protected the city from overwash damage and shoreline erosion.

Hotels, restaurants, and other businesses dependent on the tourist trade line Seawall Boulevard but are set back from the seawall by the broad (four-lane) road. Many single-family dwellings throughout Galveston are at low elevation and potential for flooding is high even though the area between Broadway Avenue and the seawall was built up after the 1900 and 1915 storms by trucking sand in. Unfortunately dune and beach sand from the island was the primary source of fill for increasing the elevation of the city (fig. 4.17); prior to its removal, this sand had protected other segments of the island.

Galveston seawall to vicinity of Eleven-Mile Road (Galveston Island: figure 4.14)

This area offers some potentially safe sites for development even though it is not protected by the seawall. Natural island

Fig. 4.17. Sand borrow pit, West Beach, Galveston Island. Source: Texas Coastal and Marine Council.

elevations range from near sea level to slightly more than 10 feet above sea level. The highest elevations are concentrated along broad, low-relief ridges that roughly parallel the long axis of the island. These beach ridges are separated by low, poorly drained swales (depressions) that sustain fresh-water marshes. Flood potential is high, and the area was submerged by 8 to 10 feet of water during Hurricane Carla. The area immediately adjacent to the beach poses several hazards including a high erosion rate (5 to 10 feet per year) and local zones of high overwash potential at sand pits. Because some of the low discontinuous dunes have been leveled or removed for fill, overwash extends 500 feet inland from and parallel to the beach. The present development is mainly single-family dwellings. The finger canals, which are common along the bay margin, are less susceptible to pollution here than elsewhere because some of the larger developments have sewer systems. Avoid buying near roads that are open to and level with the beach because they may act as overwash passes during storms.

Eleven-Mile Road to San Luis Pass (Galveston Island: figures 4.14 and 4.15)

The western one-third of Galveston Island is somewhat different from its neighboring segment in height, width, and vulnerability to storms. The island narrows progressively, and elevations are slightly lower in a westerly direction. These conditions together with the low discontinuous dunes make the area highly susceptible to flooding and overwash. The area was inundated by a storm surge of 11 feet during Hurricane Carla. In order to accommodate these storm hazards, homes in this area are constructed on stilts or the surface elevation has been artificially raised by adding material dredged from adjacent finger canals or from the dunes. Taking sand from dunes is unwise because it destroys the last vestige of natural protection. Along this barrier segment, both bay and Gulf shorelines are retreating. The Gulf shoreline is eroding at an average long-term rate of 1 to 2 feet per year, but recent rates up to 5 feet per year have been documented. The most dramatic shoreline changes have occurred near San Luis Pass where short-term erosion rates up to 70 feet per year have been recorded. Because the western tip of the island is next to a dynamic unjettied inlet, the area should be considered relatively unsafe for development. Evacuation from this island area poses an additional hazard because the shortest route (FM 3005) off the island is subject to early inundation because of low elevations.

San Luis Pass to Surfside (Follets Island: figures 4.15 and 4.18)

Follets Island is a good example of a low-profile barrier that is retreating landward. Much of the back-island area is composed of mud flats covered with marsh vegetation and numerous small ponds and tidal creeks. The dunes are very low and discontinuous, and most of the land area is below 5 feet in elevation; hence potential for overwash and flooding is very high. Several overwash channels have been active in recent times, and the island was completely inundated by an 11-foot storm surge during Hurricane Carla in 1961. The Gulf beach is eroding at rates of 5 to 10 feet per year, and the bay shoreline is also undergoing erosion at minor rates. Overall the island is a hazardous site for development, which is presently limited to a few clusters of beach cottages and a condominium complex. The beach fronting the eastern end of Follets Island adjacent to and within one mile of San Luis Pass is particularly vulnerable to major shoreline changes that occur rapidly because of shifts in the banks and channel of the tidal inlet. This same area overlies an abandoned and filled tidal inlet (Cold Pass) that was open in the late 1800s. Evacuation from Follets Island is potentially hazardous because the single route is easily flooded and washed over by storm waves.

Surfside to the Brazos River (Quintana and Bryan Beach: figures 4.18 and 4.19)

This beach segment fronts the Brazos River delta, which was actively building seaward several thousand years ago. Since that time it has been retreating landward at an impressive rate. Human activities at the mouth of the Brazos River have caused some of the most rapid and unexpected shoreline changes along the Texas coast. After the jetties were constructed in the 1890s, the shoreline built seaward as much as 6,000 feet in a 40-year span. But in 1929 the Brazos River mouth was diverted westward about 6 miles where it now empties into the Gulf. The reason this story is important to the home dweller in this area is that diverting the river cut off an important supply of beach sand and caused very rapid erosion at Surfside and Quintana on the order of 10 to 15 feet per year (fig. 4.20).

The new Brazos delta formed after the river was diverted in 1929; it reached its maximum seaward extent about 1942. Since then the delta has entered an erosional stage and is shifting its position westward toward the San Bernard River. At Bryan Beach State Park the Gulf shoreline near the river mouth is retreating at rates of 20 to 30 feet per year. Because of these extreme rates, plans for locating permanent outbuildings at the State Park were altered.

During this century the Freeport area has been inundated at least twelve times by storm surges of between 5 and 12 feet and beachfront property has been damaged or destroyed by these tropi-

72 Living with the Texas shore

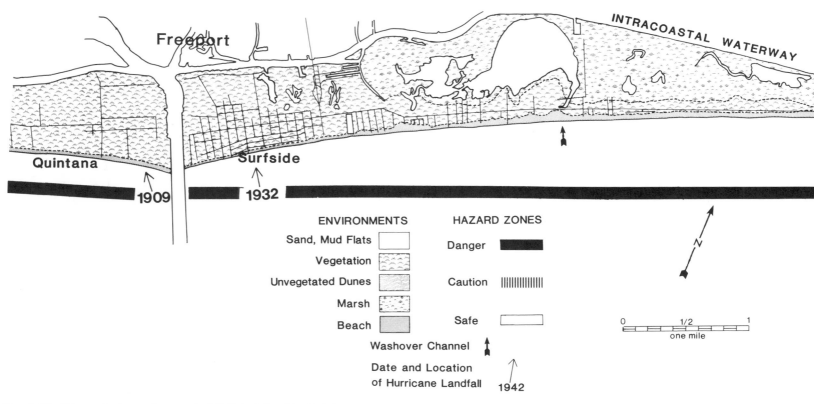

Fig. 4.18. Site analysis: eastern end of Follets Island to Quintana.

4. Selecting a site 73

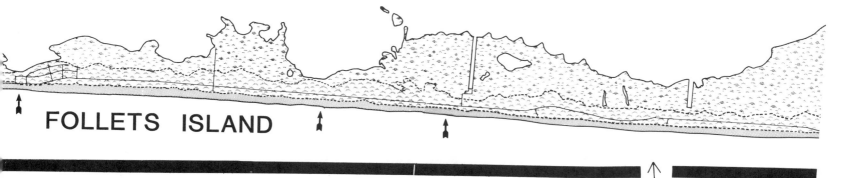

FOLLETS ISLAND

1915

- EROSION 5'-10' PER YEAR
- LOW DISCONTINUOUS DUNES
- HIGH FLOOD AND OVERWASH POTENTIAL
- NARROW ISLAND
- SINGLE EVACUATION ROUTE
- SINGLE AND MULTIFAMILY DWELLINGS

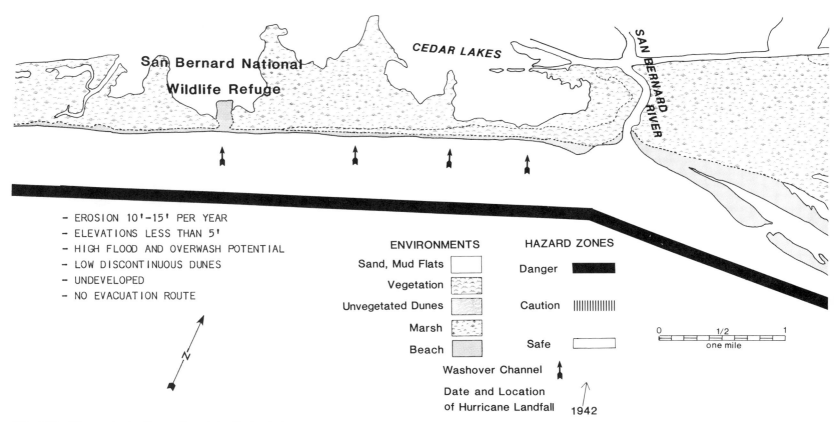

Fig. 4.19. Site analysis: Bryan Beach to Cedar Lakes.

4. Selecting a site

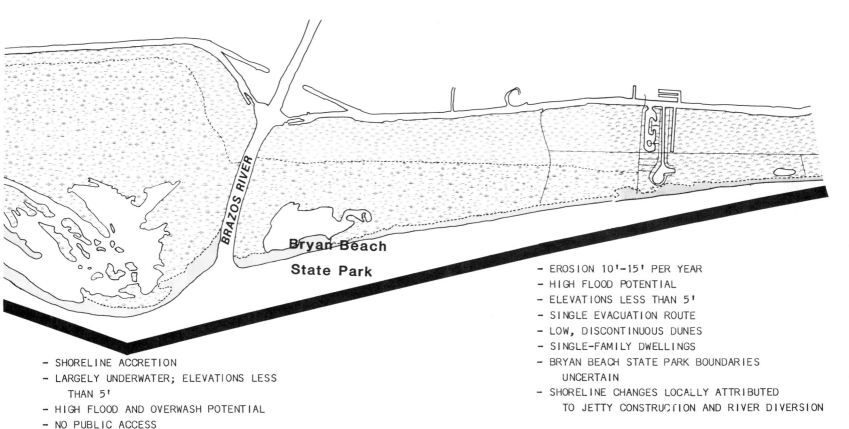

- SHORELINE ACCRETION
- LARGELY UNDERWATER; ELEVATIONS LESS THAN 5'
- HIGH FLOOD AND OVERWASH POTENTIAL
- NO PUBLIC ACCESS
- UNDEVELOPED; NO EVACUATION ROUTE

- EROSION 10'-15' PER YEAR
- HIGH FLOOD POTENTIAL
- ELEVATIONS LESS THAN 5'
- SINGLE EVACUATION ROUTE
- LOW, DISCONTINUOUS DUNES
- SINGLE-FAMILY DWELLINGS
- BRYAN BEACH STATE PARK BOUNDARIES UNCERTAIN
- SHORELINE CHANGES LOCALLY ATTRIBUTED TO JETTY CONSTRUCTION AND RIVER DIVERSION

Fig. 4.20. Beach house on an eroding beach, Surfside.

cal disturbances. But even winter storms have taken their toll along the severely eroding beach. For example in December of 1975 high waves destroyed several beach cottages near the jetties at Surfside. Other homes were undercut, exposing the foundations and destroying the septic systems.

Beach development in the vicinity of Freeport is mainly single-family dwellings built on stilts to compensate for the lack of high dunes and high elevation. The mainland location provides several evacuation routes leading inland to higher ground.

Brazos River to the San Bernard River (figure 4.19)

The western flank of the new Brazos delta is largely under water, and the sandy beach ridges that are emergent are only 1 to 2 feet above sea level; there are no permanent dunes along the narrow beach. At present this shoreline is not eroding, but because sediment supply is limited, it is very likely that it will begin to erode within the next decade. There are no roads for public access to or evacuation from this area and it is undeveloped. The extremely high potential for flooding and overwash and the instability of the shoreline make the area a dangerous one to develop in the future.

San Bernard River to Cedar Lakes (figure 4.19)

This exceptionally hazardous area is accessible only by four-wheel-drive vehicles using the beach. The poorly developed dunes are low and sparsely vegetated. Potential for flooding is high because of the low elevations (generally less than 5 feet), and several overwash channels are present. Washover deposits formed by Hurricane Carla are still visible on aerial photographs. The beach is narrow (50 to 70 feet wide) and it has been eroding at average rates of 10 to 15 feet per year over 125 years; recent rates have been as high as 30 feet per year. The area is undeveloped except for oil and gas production facilities.

Cedar Lakes to Sargent Beach (figures 4.19 and 4.21)

This shoreline segment is characterized by a narrow beach (50 feet wide) composed mostly of very shelly sand that overlies marsh mud. The mud substratum is usually exposed at low tide or after high wave activity (fig. 4.22). Elevations are generally less than 5 feet, and the area is flooded and washed over even by minor storms. The long-term erosion rate of 10 to 15 feet per year is one of the highest along the coast. Even more dramatic are the recent erosion rates which have exceeded 40 feet per year in some areas. The development, consisting of beach cottages on stilts at Sargent Beach, has lost a number of houses as a result of shoreline retreat. At least one entire row of houses and beach lots has been lost to the Gulf. Several seawalls were constructed by private interests to mitigate the shoreline retreat but they have been completely destroyed. Evacuation from the area could be hazardous because the single evacuation route (FM 457) has low elevations and a high potential for flooding. The road was under about 8 to 10 feet of water during Hurricane Carla in 1961.

Sargent Beach to Brown Cedar Cut (figure 4.21)

The 5-mile stretch of beach westward of the access road (FM 457) is narrow and relatively steep; it contains abundant shells and overlies marsh mud. These conditions make for treacherous driving even in the four-wheel-drive vehicles that are required to reach this area. The dunes are low, discontinuous, and sparsely vegetated, and the beach is eroding at 10 to 15 feet per year. Elevations are mostly less than 5 feet, and the area is susceptible to flooding and overwash. Several washover channels were opened in Hurricane Carla in 1961, and these covered the land with 10 feet of water. The narrow barrier is largely undeveloped except for a few fishing camps clustered near Brown Cedar Cut and along the bay shoreline. Brown Cedar Cut is a small tidal inlet that shoals (fills with sand) periodically but is open most of the time and almost always after a storm. The inlet is relatively stable in its present position, although it migrated 1,400 feet southwestward between 1937 and 1953.

East Matagorda Peninsula to Spring Bayou (figures 4.23 and 4.24)

Matagorda Peninsula is typical of retreating or low-profile barriers. It consists of a narrow beach backed by a narrow grassflat constructed on overwash sands. Island width increases westward, but much of the increased back-island area is occupied by marsh that is mainly submerged. The line of dunes is low, discontinuous, and narrow. In places, dunes are absent and the vegetated flat is actually an overwash terrace. Elevations are mostly less than 5 feet, so potential for flooding and overwash is high. Over 30 washover channels have been mapped along this beach segment. Not only are washover channels numerous, but the Gulf beach is eroding at rates of 5 to 10 feet per year. The bay shoreline is also eroding at rates between 2 and 5 feet per year. Four-wheel-drive vehicles are required to reach this area, which is presently undevel-

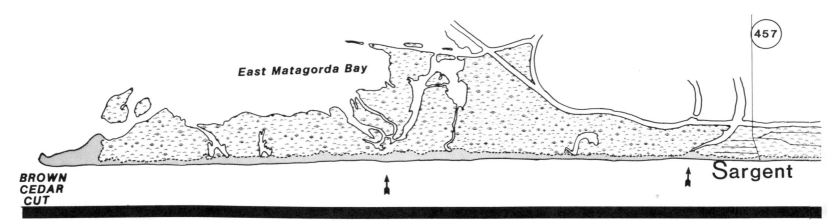

- EROSION 10'-15' PER YEAR
- BAY SHORELINE EROSION
- LOW DISCONTINUOUS DUNES
- ELEVATIONS LESS THAN 5'
- HIGH FLOOD AND OVERWASH POTENTIAL
- UNDEVELOPED
- NO EVACUATION ROUTE

Fig. 4.21. Site analysis: Cedar Lakes to Brown Cedar Cut.

4. Selecting a site 79

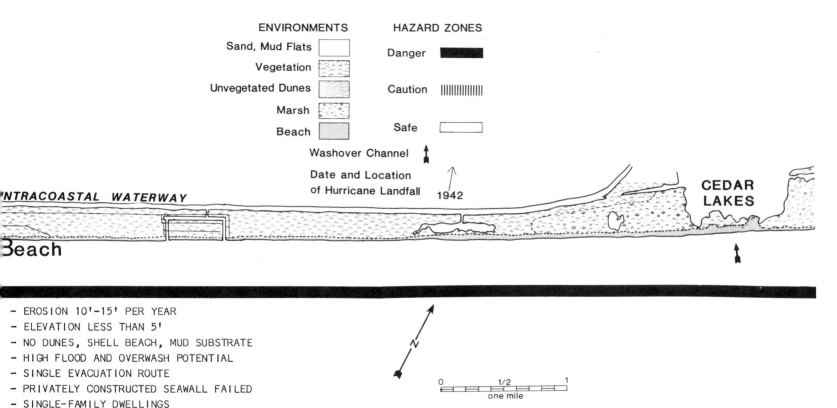

- EROSION 10'-15' PER YEAR
- ELEVATION LESS THAN 5'
- NO DUNES, SHELL BEACH, MUD SUBSTRATE
- HIGH FLOOD AND OVERWASH POTENTIAL
- SINGLE EVACUATION ROUTE
- PRIVATELY CONSTRUCTED SEAWALL FAILED
- SINGLE-FAMILY DWELLINGS

Fig. 4.22. Outcrop of marsh mud, Sargent Beach.

oped except for oil and gas production facilities. The high rate of erosion, narrow width, low elevation, and resulting widespread overwash make this island unsuitable for development.

Spring Bayou to the Colorado River (figure 4.24)

Conditions along this short barrier segment are similar to those on the remainder of East Matagorda Peninsula except the dunes are better developed, elevations are slightly higher, and the barrier is slightly wider. Even though conditions are somewhat safer for development, the area is still susceptible to flooding and overwash, especially near Spring Bayou, which is a permanent overwash channel. These washovers were active and the area was inundated by an 11-foot storm surge during Hurricane Carla. In addition to these hazards the beach is eroding from 5 to 10 feet per year.

Both condominiums and single-family dwellings are clustered near the mouth of the Colorado River. At present, development is light but it is increasing rapidly. Storm evacuation could be a problem because the single access road (FM 2031), which parallels the river, is easily flooded and crosses a drawbridge at the Intracoastal Canal.

West Matagorda Peninsula to Greens Bayou (figures 4.24 and 4.25)

Matagorda Peninsula west of the Colorado River is used primarily as a privately owned cattle ranch with no public access. As such, it is undeveloped except for a few outbuildings and private homes. This is a dynamic island that is migrating landward by beach erosion and storm overwash processes. The Gulf shoreline is retreating at 5 to 10 feet per year, and during major storms such as Carla in 1961, overwash fans deposited in the back-barrier marsh caused the bay shoreline to accrete or build into the bay. The bay shoreline between washover fans, however, is also eroding at rates of 5 to 10 feet per year. Dunes along this island segment are low and discontinuous, elevations are generally less than 10 feet, and the area has a high potential for flooding. It also suffered the most severe damage during Hurricane Carla, which severed the island with numerous deep washover channels (fig. 2.7) and covered the land with a 12-foot storm surge. A short stretch just west of the Colorado River is afforded some protection by dunes

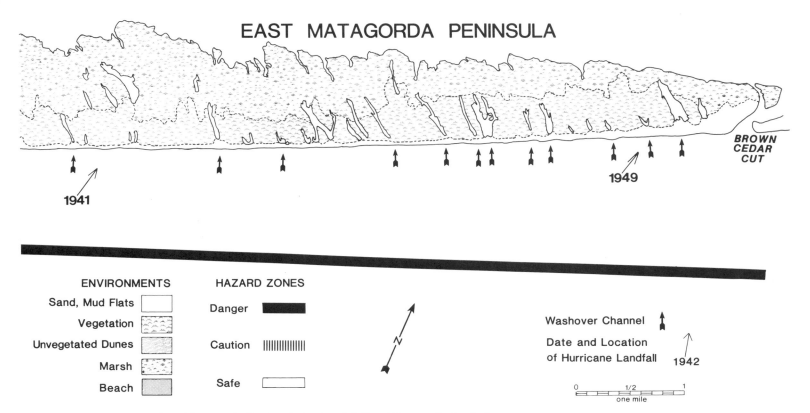

Fig. 4.23. Site analysis: East Matagorda Peninsula in the vicinity of Brown Cedar Cut.

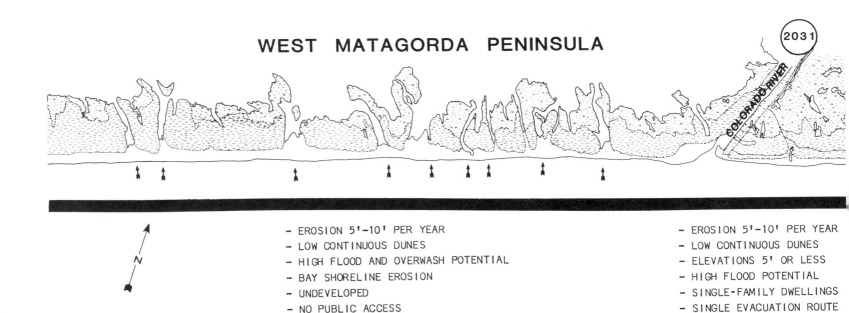

Fig. 4.24. Site analysis: East and West Matagorda Peninsula in the vicinity of the Colorado River.

4. Selecting a site 83

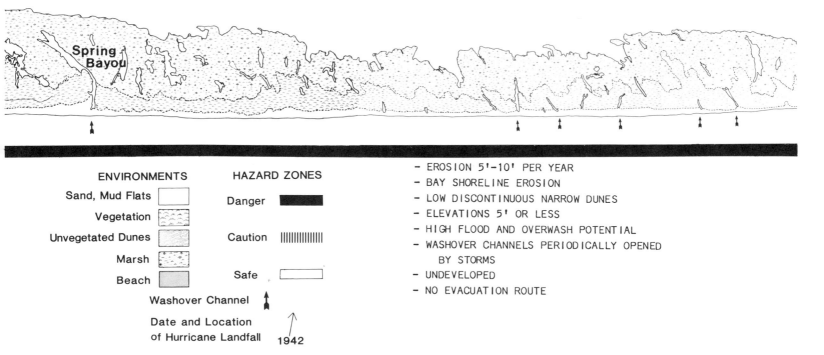

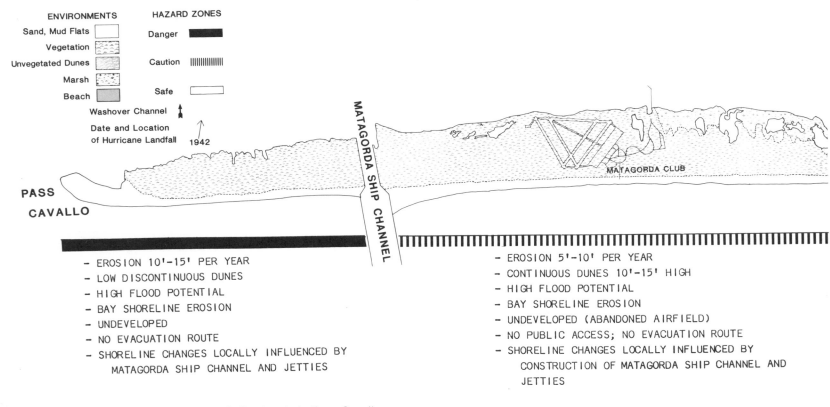

Fig. 4.25. Site analysis: West Matagorda Peninsula to Pass Cavallo.

4. Selecting a site

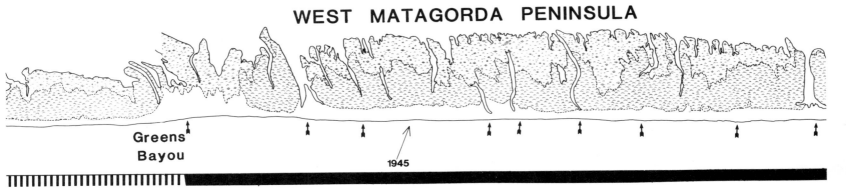

WEST MATAGORDA PENINSULA

Greens Bayou

1945

- EROSION 5'-10' PER YEAR
- HIGH FLOOD AND OVERWASH POTENTIAL
- WASHOVERS PERIODICALLY OPENED BY STORMS
- BAY SHORELINE EROSION
- UNDEVELOPED
- NO PUBLIC ACCESS
- NO EVACUATION ROUTE

N

0   1   2
    miles

locally supplied by sand from the river mouth. In its natural state, the Colorado River emptied into Matagorda Bay, but the channel was dredged across Matagorda Peninsula in 1936. Even though the river now empties into the Gulf, the additional sand supply has been insufficient to alter the rate of shoreline retreat. The combined threats of shoreline erosion, numerous washover channels, and storm flooding make this area unsuitable for development.

Greens Bayou to Matagorda Ship Channel (Matagorda Peninsula: figure 4.25)

Greens Bayou is a broad overwash area that occupies the position of a former tidal inlet. As such, it is easily flooded and the site of strong currents during storms. Most of the back-island area is also flood-prone since elevations are generally less than 5 feet. In contrast to other nearby areas, this barrier segment has a line of continuous dunes 10 to 15 feet high with maximum elevations on the order of 25 feet; hence the potential for overwash landward of the beach is minimal. Although overwash is not a serious threat in this area, scour from storm waves can be substantial as shown by Hurricane Carla. During this storm the low dunes and vegetation line eroded nearly 400 feet.

Bay shoreline erosion for this stretch is low, about 2 feet per year, and average erosion rates for the Gulf shoreline are also low, between 0 and 5 feet per year. However, some local shoreline changes have occurred due to the jetty construction for Matagorda Ship Channel, which was dredged across the island in 1963. Shortly thereafter, the beach grew seaward east of the jetties, and substantially increased erosion was experienced west of the jetties.

The Matagorda Ship Channel design was based on careful model studies made at the Corps of Engineers facilities in Vicksburg, Mississippi. As so often happens, nature refused to behave as she was supposed to. The major difficulty was erosion on the channel banks by tidal currents going in and out of Matagorda Bay. Considerable unexpected costs resulted when bulkheads were constructed to protect the channel banks. A small abandoned airbase from the Second World War is the only development in this area which is inaccessible to the public.

Matagorda Ship Channel to Pass Cavallo (Matagorda Peninsula: figure 4.25)

This short barrier segment is undeveloped and there is no public access. The dunes are moderately well developed near the ship channel, and elevations have been artificially increased with dredged material, but dunes are low and discontinuous and elevations decrease westward. Overall, the potential for flooding is high and the western tip of the peninsula is overwashed periodically. The erosion rate of the Gulf shoreline is about 10 to 15 feet per year, and the bay shoreline is retreating at about 5 feet per year. This higher than normal rate is probably related to the jetties at Matagorda Ship Channel. The sand eroded west of the jetties apparently caused a spit to extend westward 2,800 feet into Pass Cavallo. The rapid growth of the spit occurred in the nine-year period between 1965 and 1974. The high potential for flood and overwash, the shoreline erosion, and the instability of the inlet make this area unsafe for development.

Matagorda and San Jose Islands (figures 4.26, 4.27, 4.28, and 4.29)

Matagorda and San Jose Islands are high-profile barriers that are characterized by wide sand beaches, high continuous and stable dunes, broad vegetated flats, and fringing marshes and ponds along the bay margin. The back-island areas with elevations less than 5 feet have been flooded at least twelve times during this century. Although the back-island marshes and the low swales (depressions) of the barrier flat are susceptible to flooding, the continuous dune ridge provides ample protection from storm overwash. Except for the immediate shoreline and the active or potentially active dune areas, these islands afford numerous safe sites for development.

Recently the eastern two-thirds of Matagorda Island was declared surplus government property. Prior to 1976 this area was an active Air Force base and bombing range. Its beauty and nearly pristine state would make it an ideal nature preserve and natural laboratory for studying barrier-island processes and ecosystems. But pressure is mounting to provide public access for recreational purposes in addition to hunting and fishing. In 1982 management responsibilities for Matagorda Island were transferred from the federal government to the Texas Department of Parks and Wildlife, which plans to maintain the undeveloped island in its present state.

East end of Matagorda Island (figure 4.26)

The eastern end of Matagorda Island along the shore of Pass Cavallo has been a strategic site for military outposts for more than a century. Several forts were built in this area and subsequently destroyed by hurricane flooding and southerly migration of the tidal inlet. Early attempts to construct jetties at the bay entrance were abandoned in 1888 after Indianola was destroyed. Today the only remnants of a once populated area are partly filled Civil War trenches, an abandoned lighthouse, and an abandoned Air Force base. The beach adjacent to Pass Cavallo is subject to rapid and erratic shoreline changes. The tendency for southerly migration of the pass and attendant beach erosion on Matagorda Island could be aggravated by the spit building westward from Matagorda Peninsula. Hence, a zone within 0.5 mile of the inlet probably should be avoided. Elsewhere the dunes are well established and the island is wide and relatively safe for development.

Air Force Base to Power Lake vicinity (Matagorda Island: figures 4.26 and 4.27)

Over the last 100 years this stretch of Matagorda Island has grown slightly seaward. During the last 10 years or so, however, the beach has been eroding a few feet per year, a relatively slow rate by Texas standards. The broad sand beach is backed by a line of continuous dunes that are 15 to 25 feet high. Furthermore, the island is relatively wide. The fact that Matagorda Island Air Force Base has lasted without major loss since the Second World War is

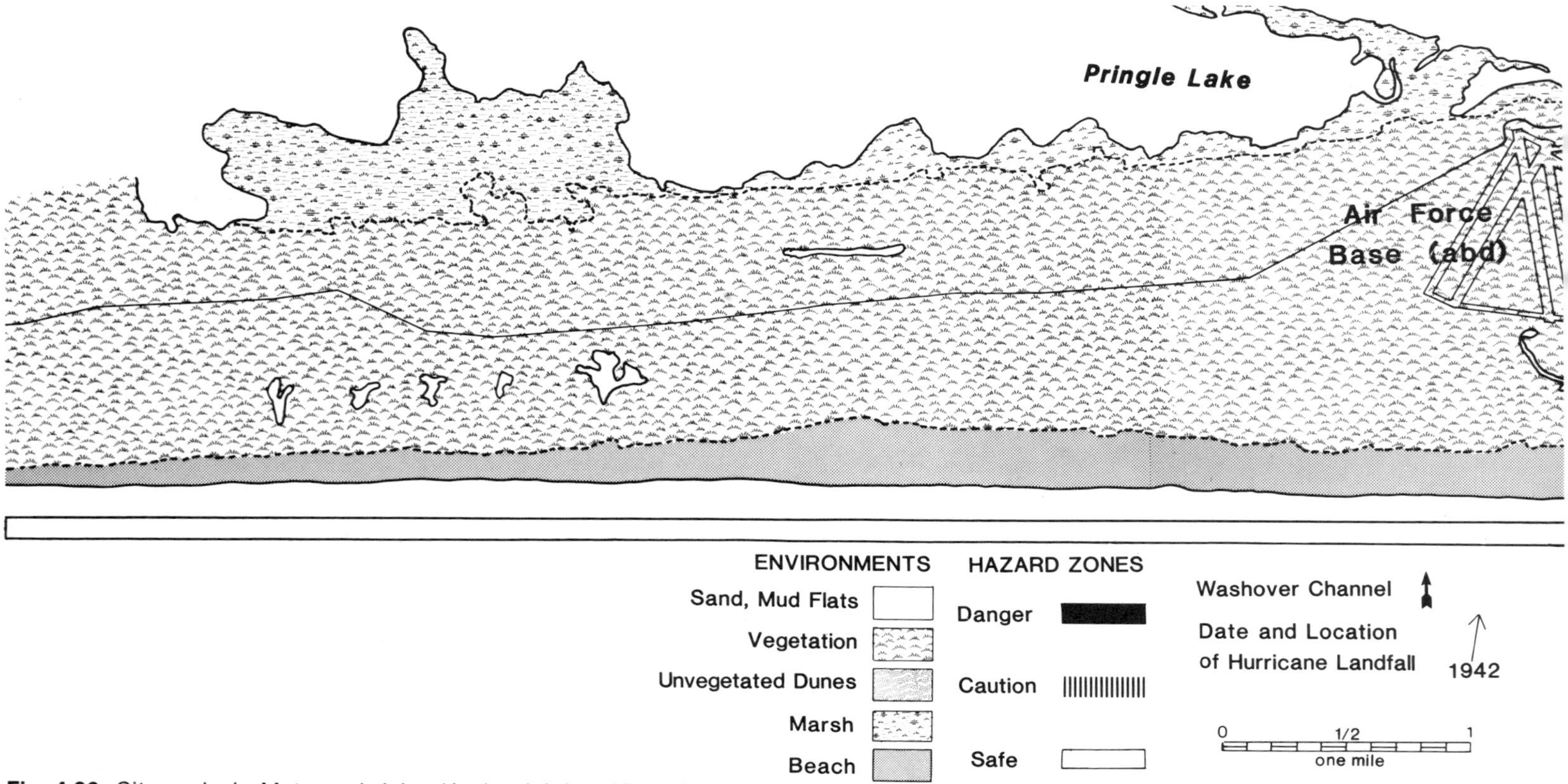

Fig. 4.26. Site analysis: Matagorda Island in the vicinity of Pass Cavallo.

4. Selecting a site 89

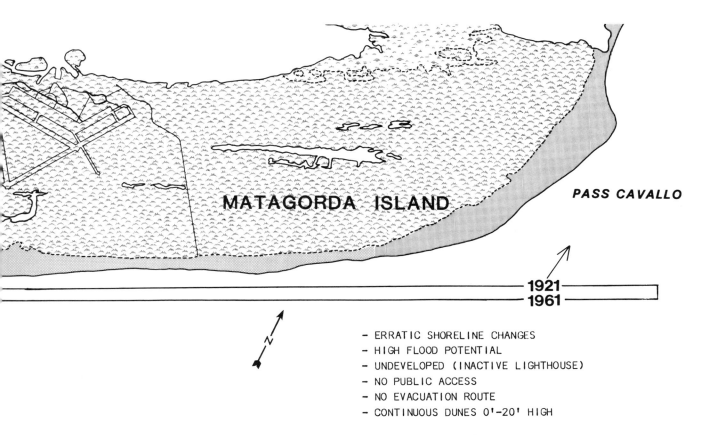

- ERRATIC SHORELINE CHANGES
- HIGH FLOOD POTENTIAL
- UNDEVELOPED (INACTIVE LIGHTHOUSE)
- NO PUBLIC ACCESS
- NO EVACUATION ROUTE
- CONTINUOUS DUNES 0'-20' HIGH

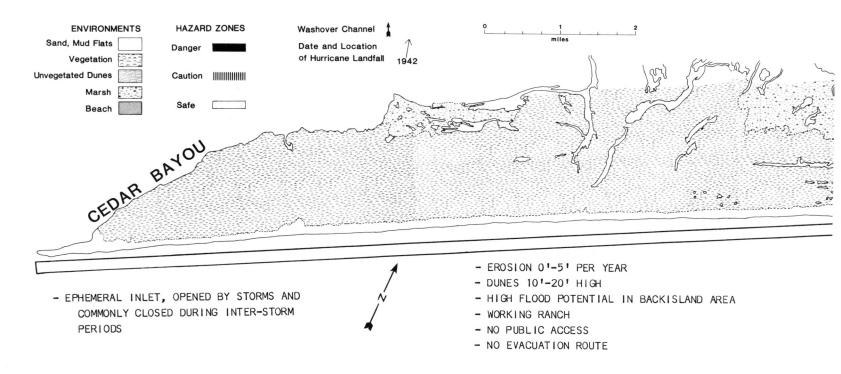

Fig. 4.27. Site analysis: Matagorda Island from Power Lake to Cedar Bayou.

4. Selecting a site 91

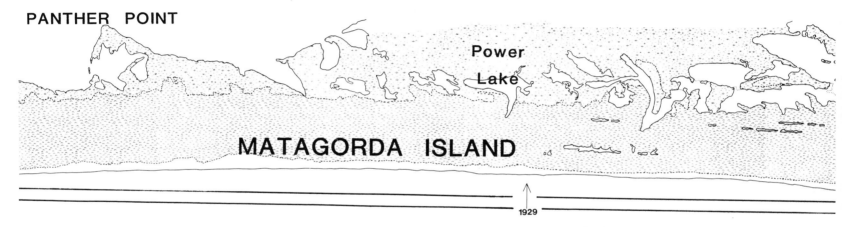

PANTHER POINT

Power Lake

MATAGORDA ISLAND

1929

- EROSION 0'-5' PER YEAR
- ACTIVE DUNES NEAR BEACH
- HIGH FLOOD POTENTIAL IN BACKISLAND
- UNDEVELOPED (INACTIVE BOMBING RANGE)
- NO PUBLIC ACCESS
- NO EVACUATION ROUTE

- EROSION 0'-5' PER YEAR
- HIGH FLOOD POTENTIAL IN BACKISLAND
- CONTINUOUS DUNES 15'-25' HIGH
- NO PUBLIC ACCESS
- NO EVACUATION ROUTE
- UNDEVELOPED (INACTIVE BOMBING RANGE AND AIR FORCE BASE)

a sure indication of the relative safety of this island for development. The base survived Hurricane Carla, the eye of which passed 5 miles to the east, up Pass Cavallo. In this area the bay shore is undergoing minor erosion and back-island marshes are subject to flooding because of low elevation. There is no public access to this area, no evacuation route, and no private development.

Power Lake vicinity to Panther Point vicinity (Matagorda Island: figure 4.27)

This shoreline stretch is characterized by a band of active sand dunes up to one-half mile in width adjacent to the Gulf beach. This could create a problem for development, but most of this barrier segment is wide and high. The bayshore is eroding slightly, and recently the Gulf beach has been retreating at less than 5 feet per year. The island is wide, however, and there is no present danger from shoreline erosion. Similarly, overwash is not likely in this area because the beach is several hundred feet wide and the continuous line of dunes is 15 to 20 feet high. However blowouts (migrating dunes) were numerous and back-island dunefields were active in the 1930s because of the loss of vegetation caused by hurricane flooding coupled with severe droughts. The dunes have since become re-established during periods of ample rainfall.

This island segment is uninhabited and undeveloped with no access or evacuation route. Wild game such as deer, turkey, coyotes, and rabbits abound on this island.

Panther Point vicinity to Cedar Bayou (Matagorda Island: figure 4.27)

This stretch of barrier is characterized by extensive sand ridges that occur along the mid-island, parallel the Gulf shoreline. The beach ridges have elevations of 10 to 15 feet, and adjacent dune elevations range from 10 to 20 feet with a maximum elevation of 50 feet. At present, the shoreline is eroding at 0 to 5 feet per year. If one looks at the 100-year shoreline trend, however, he will see that the beach has actually grown seaward. The erosion figure of 0 to 5 feet per year refers to the last decade only. Although flooding along the central Texas coast was extensive during Hurricane Carla in 1961, much of the western third of Matagorda Island was above flood levels. This area is presently a privately owned cattle ranch with no public access and no land evacuation route.

Cedar Bayou to North Pass (San Jose Island: figures 4.28 and 4.29)

Matagorda and San Jose Islands are separated by Cedar Bayou, a minor tidal inlet that tends to shoal (fill with sand) at the Gulf beach when storms are infrequent. As with most small inlets, Cedar Bayou has followed a pattern of breaching, southwesterly migration, and shoaling. These changes in position are limited to the vicinity of the Gulf beach, and the main channel between the islands is stable; however, it does serve as a washover area during major storms.

4. Selecting a site 93

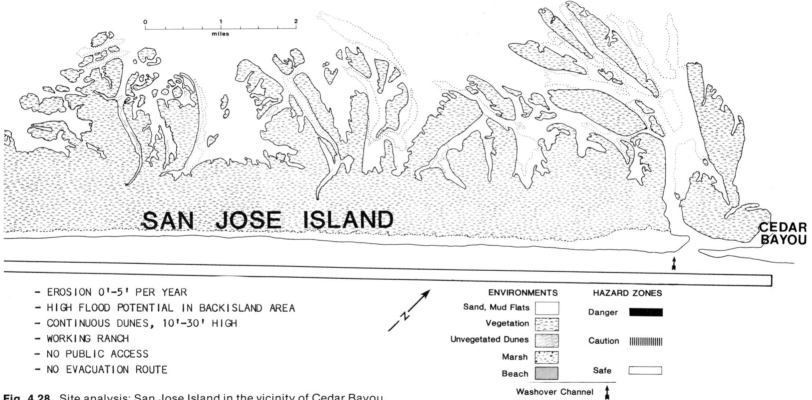

Fig. 4.28. Site analysis: San Jose Island in the vicinity of Cedar Bayou.

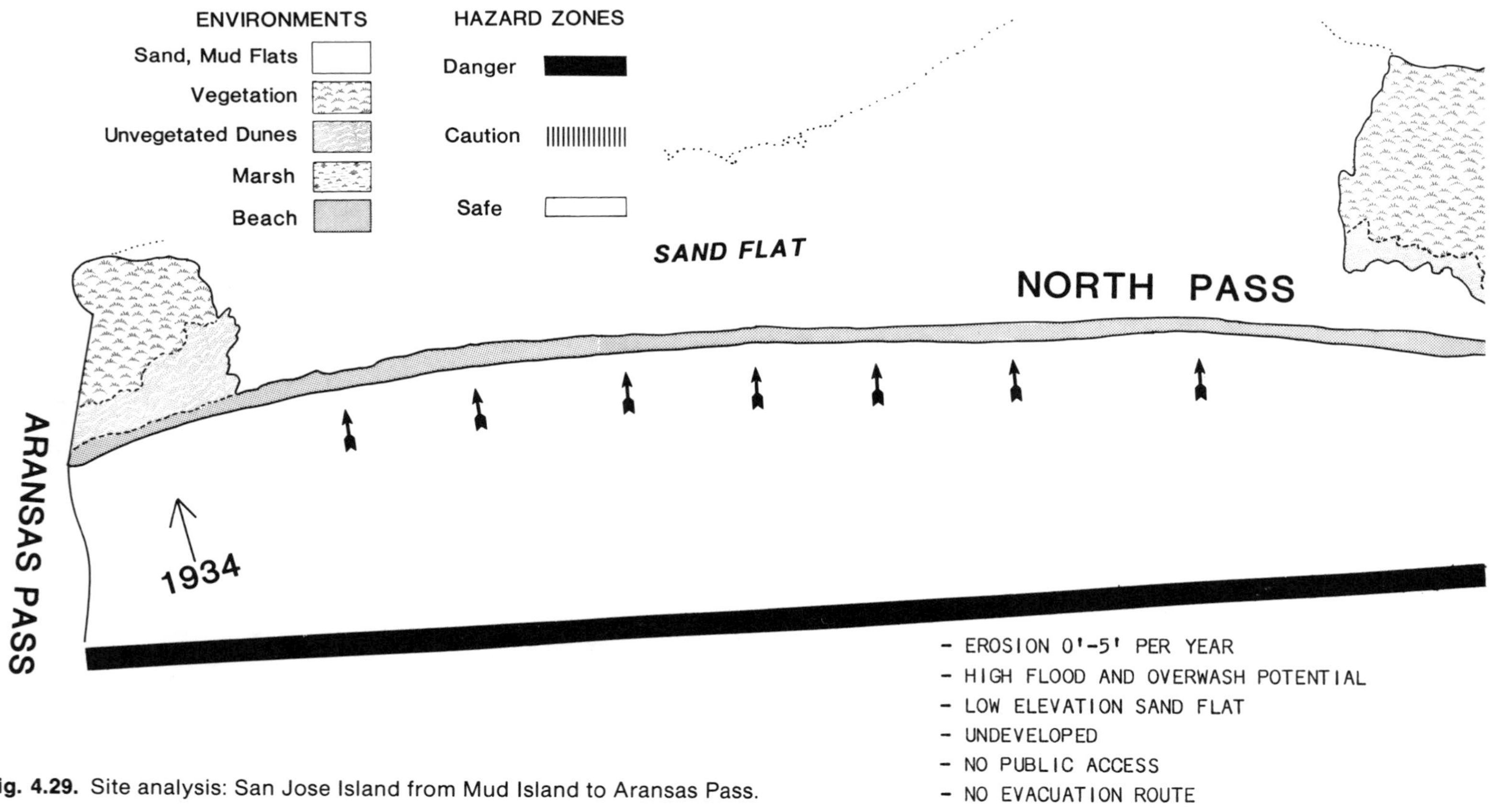

94 Living with the Texas shore

Fig. 4.29. Site analysis: San Jose Island from Mud Island to Aransas Pass.

ENVIRONMENTS
- Sand, Mud Flats
- Vegetation
- Unvegetated Dunes
- Marsh
- Beach

HAZARD ZONES
- Danger
- Caution
- Safe

SAND FLAT
NORTH PASS
ARANSAS PASS
1934

- EROSION 0'–5' PER YEAR
- HIGH FLOOD AND OVERWASH POTENTIAL
- LOW ELEVATION SAND FLAT
- UNDEVELOPED
- NO PUBLIC ACCESS
- NO EVACUATION ROUTE

4. Selecting a site 95

Mud Island

SAN JOSE ISLAND

Washover Channel

Date and Location
of Hurricane Landfall 1942

0   1/2   1
one mile

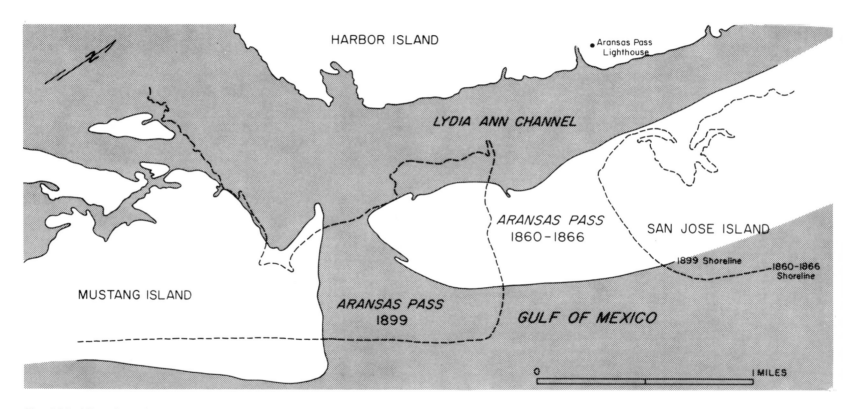

Fig. 4.30. Migration of Aransas Pass. Source: Morton and Pieper (1976). See reference 63 in appendix C.

Until recently San Jose Island was called St. Joseph's Island. On some topographic maps the island is still called St. Joseph's Island. The name will be changed as the maps are updated.

Like Matagorda Island, San Jose is marked by a broad area of sandy beach ridges and intervening swales (depressions). The ridges, which are 10 to 15 feet high, are separated from the Gulf beach by a continuous line of stable dunes 10 to 30 feet high. These dunes protect the barrier flat from overwash. For the better part of this century the Gulf beach was relatively stable but recently it has entered an erosional phase and is retreating at a rate of a few feet per year. The bay shoreline is also eroding at rates of a few feet per year. Much of the bay margin of this island is marsh that is low in elevation and highly susceptible to flooding.

The island is undeveloped except for a palatial summer home with an airstrip and outbuildings that support the working cattle ranch. There is no public access to this barrier and no storm evacuation route by land.

North Pass to Aransas Pass (San Jose Island: figure 4.29)

The southern tip of San Jose Island is a very low, unvegetated sand flat with low sand hummocks (mounds) that formed during the middle to late 1800s. During this period, which was prior to jetty construction, Aransas Pass was unstable, and southwesterly migration rates up to 260 feet per year caused erosion of Mustang Island and attendant deposition on San Jose Island (fig. 4.30). The lighthouse on Harbor Island behind San Jose once shone its beacon straight through the inlet in the 1850s. Now the inlet is located over a mile southwest and the lighthouse no longer functions.

The North Pass is extremely hazardous. Not only is the Gulf shoreline eroding at 0 to 5 feet per year, but the entire area is overwashed frequently because elevations are mostly less than 5 feet. The small area of sand dunes near Aransas Pass is largely formed from dredged material.

Aransas Pass to Mustang Island State Park (Mustang Island: figure 4.31)

Engineering modifications at Aransas Pass were first made in 1868 in order to stop rapid migration of the inlet (fig. 4.30). The first attempts failed within 3 years, and after several other attempts at jetty construction the present pair of jetties was completed about 1910. In the immediate vicinity (within 1 mile) of the jetties the shoreline has built seaward leaving a broad sand flat between the dunes and the beach. This has been a commonly observed phenomenon wherever natural Texas inlets are jettied.

Erosion of the Gulf beach on Mustang Island is variable and ranges from 1 to 2 feet per year. The bay shoreline is also eroding at a comparable rate except for the western end in the vicinity of the state park. This stretch has actually built landward by dunes migrating into Laguna Madre. The back-island area is low in elevation and is subject to both flooding and shoreline erosion. Storms have flooded this area at least fourteen times during this

98 Living with the Texas shore

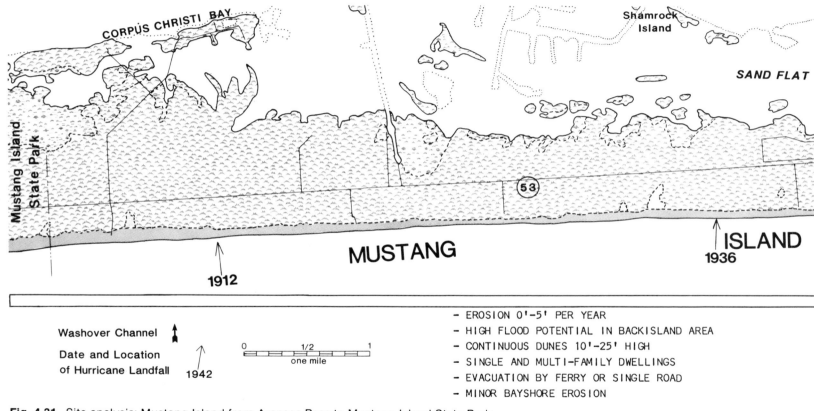

Fig. 4.31. Site analysis: Mustang Island from Aransas Pass to Mustang Island State Park.

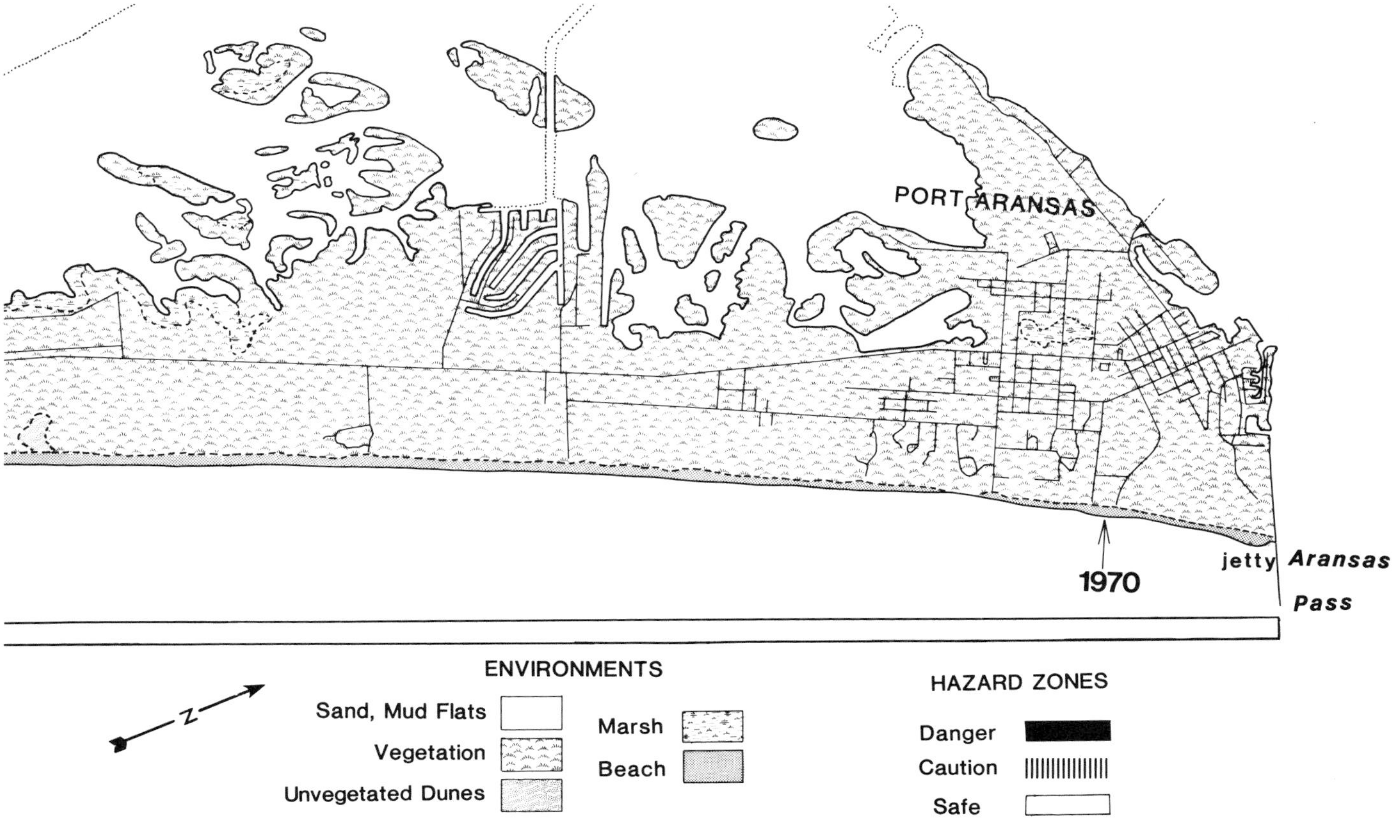

4. Selecting a site   99

PORT ARANSAS

1970

jetty *Aransas*
Pass

ENVIRONMENTS

Sand, Mud Flats
Vegetation
Unvegetated Dunes
Marsh
Beach

HAZARD ZONES

Danger
Caution
Safe

century. Surge heights of 8 feet or more were recorded in 1916, 1919, 1933, 1934, 1942, 1961, 1967, and 1970.

The island is generally well protected from overwash by the continuous dunes that are 10 to 25 feet high. When it was originally established as a fishing village and demarcation point for ships' pilots, Port Aransas occupied the highest ground on the island—behind the high dune ridge. For the most part, developers have respected the higher elevations and natural vegetative cover provided by the dunes. As a result, most of the condominiums have been constructed behind the first row of dunes (fig. 4.32).

Some landowners have placed sand fences in front of the dunes to trap blowing sand. The success of these dune building projects depends largely on the storm frequency over the next decade because newly created dunes are most vulnerable to destruction during the first few years before they are stabilized by vegetation. Even longevity, however, is no guarantee that dunes will survive a major storm unscathed. On the contrary, the high fore-island dunes along Mustang Island and other Texas barriers have been severely eroded by extreme storms. In fact, in 1961 Carla caused dune erosion of 200 feet. This coastline exemplifies numerous stages of development, ranging from high-rise condominiums at Port Aransas to natural environments midway on the island between beach access roads. Most of the island, however, is privately owned, and extensive development is planned. The present development is concentrated in the vicinity of Port Aransas and consists of single- and multifamily dwellings and commercial

Fig. 4.32. Condominiums on Mustang Island.

establishments with some trailer courts (fig. 4.33) that depend on the heavy tourist population. Storm evacuation from Mustang poses a potential problem. Because of time delays and their small size the ferries from Port Aransas to Harbor Island cannot be depended upon to evacuate significant numbers of people. The John F. Kennedy Causeway to Corpus Christi is very low over

4. Selecting a site   101

Fig. 4.33. Port Aransas trailer park after Hurricane Celia (1970). Source: Texas Coastal and Marine Council.

much of its extent across Laguna Madre. It can be flooded during the early stages of a major storm.

Finger canal development on Mustang Island is not extensive, and most of the island is on a sewage system; hence, the chances are good that the canals will not become highly polluted.

Corpus Christi Pass to Packery Channel (Mustang Island: figure 4.34)

Three passes (Corpus Christi Pass, Newport Pass, and Packery Channel) located within a 4-mile segment of southern Mustang Island served as natural tidal inlets in the late 1800s. This area of inlet migration once separated Mustang and North Padre Island, but the inlets closed after the Corpus Christi ship channel was dredged and Aransas Pass was enlarged. This barrier stretch has low elevations, and the dunes are low and discontinuous. Flooding is frequent and overwash occurs even when storm winds are from the north. Extensive flooding occurred from a storm surge of 9 feet during Hurricane Beulah. During such storms the passes are briefly reopened. The area poses potential problems for storm evacuation because in the past approaches for the bridges over these washover channels have washed out, completely blocking any exit from the island.

Condominiums and motels were built on the Gulf beach in front of the dunes, so a large seawall was constructed to provide storm protection. The seawall was severely damaged in 1980 during Hurricane Allen as were all seawalls (see fig. 4.35). These structures have caused the beach to narrow and steepen. Since the Gulf beach is retreating 0 to 5 feet per year, it will not be long before either (1) the beach will be gone and the development will be at the water's edge or (2) a beach replenishment project will be carried out, partially funded by local citizens.

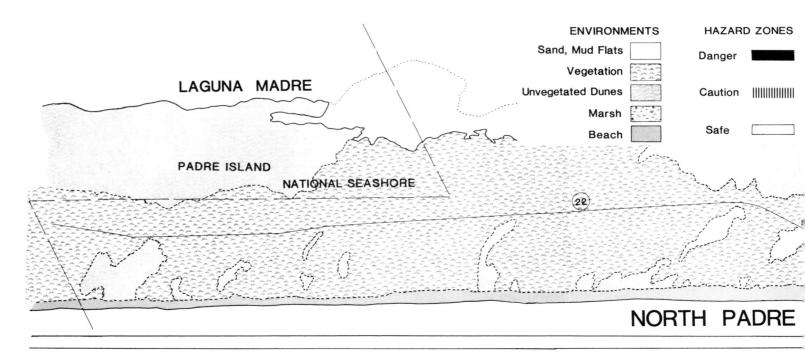

Fig. 4.34. Site analysis: Mustang Island and North Padre Island from Corpus Christi Pass to Padre Island National Seashore.

4. Selecting a site   103

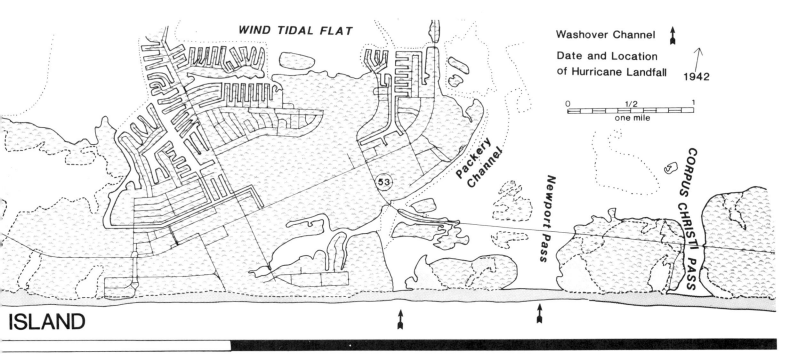

- EROSION 0'-5' PER YEAR
- HIGH FLOOD AND WASHOVER POTENTIAL
- LOW, DISCONTINUOUS DUNES BETWEEN OVERWASH CHANNELS
- SINGLE AND MULTIFAMILY DWELLINGS
- SINGLE EVACUATION ROUTE
- AVOID BEACH AND OVERWASH AREAS.

Unfortunately, local citizens who were not responsible for the development will suffer the consequences, either through increased taxes or through loss of their recreational beach. Development here is both single- and multifamily dwellings. Finger canals excavated off Packery Channel might be particularly dangerous during the next major storm.

Packery Channel to Padre Island National Seashore (Padre Island: figure 4.34)

North Padre and Mustang Islands were separated by Packery Channel, which migrated throughout the late 1800s from the area of Corpus Christi Pass to its final position. According to local authorities the channel ceased to function as a natural tidal inlet after Aransas Pass was deepened and an artificial channel was dredged into Corpus Christi Bay between Harbor Island and Mustang Island. Today, Packery Channel is being modified to serve as a waterway for boats between Padre Isles development and Laguna Madre. The channel got its name from the meat packeries that were located on the inlet in the 1870s to handle the oversupply of cattle from ranching operations. Overgrazing by cattle has been cited as one of the reasons that active dunes are so prevalent on North Padre. Extensive fields of dunes 5 to 15 feet high are migrating to the northeast at rates of up to 75 feet per year.

Beach development across the artificially filled inlet should be considered hazardous because the buildings lack protection from

Fig. 4.35. Seawall on North Padre Island that failed during Hurricane Allen in 1980. Photograph by Bill White.

storm waves and flooding. Massive seawalls in this area were damaged in 1980 by waves from Hurricane Allen even though the storm crossed the coast more than 80 miles to the south (fig. 4.35). Even under normal conditions the beach is retreating a few feet each year. North Padre is an island of contrasts. Some very well-built homes are on dangerous building sites. In an emergency, even the residents of well-constructed homes on many safe sites in the back-island area must exit the island on potentially hazardous escape routes. Since most development on North Padre has occurred relatively recently, there has not been sufficient time to see how it will respond to the forces of nature.

Mansfield Channel to Veterans Park (South Padre Island: figures 4.36, 4.37, and 4.38)

Mansfield Channel was dredged across Padre Island in 1962 to allow ships and small boats to pass between Port Mansfield and the Gulf. The small jetties at the mouth of the channel are causing erosion to the north and deposition of sand to the south because the dominant direction of sand movement is from south to north.

This stretch of Padre Island is undeveloped and has a character that is different from any other privately owned Texas barrier island. Patches of sand dunes with maximum elevations of 25 feet are separated by numerous broad and very well-defined washover channels. At least 60 washovers were opened during back-to-back storms in 1933, and an equal number were reported following Hurricanes Beulah (1967) and Allen (1980). The Gulf beach is receding at rates of 10 to 15 feet per year. Dunes are generally discontinuous and do not occur in distinct bands. The dunes retreated 50 to 350 feet during Hurricane Beulah with the greatest erosion occurring along the margins of washovers. Many of the dune fields are active, especially north of the Willacy-Cameron county line.

Migrating sand dunes north of the development are less of a problem here than elsewhere, but periodically they still cause sand to drift across the main highway—sand that is scraped up, trucked off, and used to fill low areas.

The back-barrier flats are very low in elevation and flood readily in even minor storms. From Mansfield Channel all the way to the Brazos Santiago Pass, the island is migrating in the classical way—erosion on the Gulf side and deposition in Laguna Madre. All development plans should take this into account. Access to and evacuation from this hazardous area is by four-wheel-drive vehicle, and although it is undeveloped, it has been subdivided and lots are being sold.

Veterans Park to Brazos Santiago Pass (South Padre Island: figure 4.39)

Resort development with the greatest density is found on South Padre Island where high-rise condominiums line the beachfront. Single-family dwellings intermixed with condominiums typify the development (fig. 4.40). In the future most of the Gulf-side development will likely be multifamily structures. Land-use practices on South Padre Island have not been good. Significant volumes of sand have been removed to increase the elevation of individual building sites. These notches in the dunes may create artificial overwash passes which could endanger back-island areas. Erosion rate of the Gulf beach is 5 to 10 feet per year. Natural sand transport is from south to north on South Padre Island, thus the

106   Living with the Texas shore

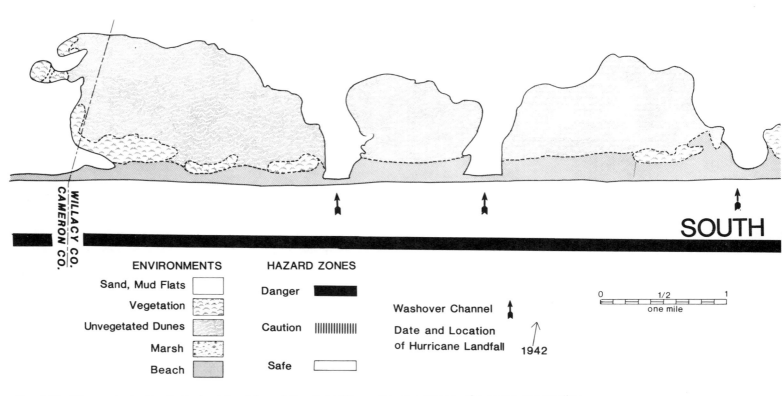

Fig. 4.36. Site analysis: South Padre Island from Mansfield Channel to the Willacy-Cameron county line.

4. Selecting a site 107

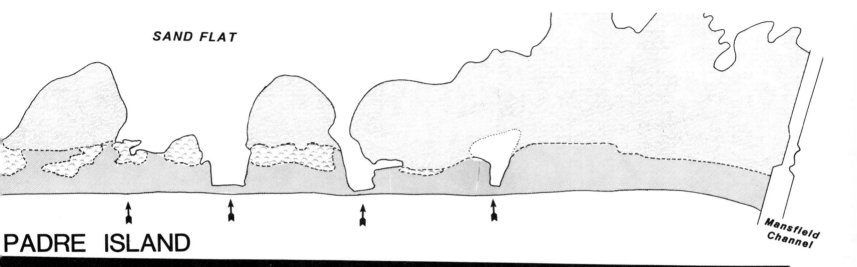

PADRE ISLAND

- EROSION 5'-10' PER YEAR
- ACTIVE DUNE FIELDS
- HIGH FLOOD AND WASHOVER POTENTIAL
- WASHOVERS PERIODICALLY OPENED BY STORMS
- UNDEVELOPED

108 Living with the Texas shore

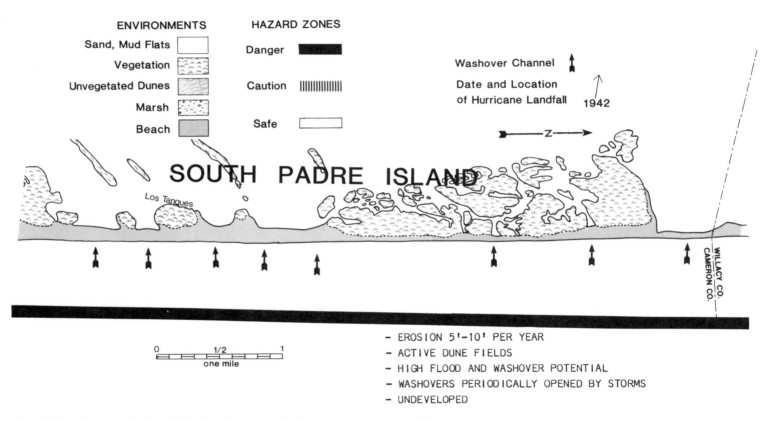

Fig. 4.37. Site analysis: South Padre Island near Willacy-Cameron county line.

jetties at Brazos Santiago Pass contributed to the erosion problem by trapping sand that should have gone to South Padre Island. The erosion rate of this narrow segment of the island (1,000 feet wide) decreases slightly to the south as one gets closer to Isla Blanca Park and the jetties. Much of the developed lagoon side of South Padre is bulkheaded. In its natural state this island stretch was migrating landward with erosion on the Gulf shoreline and deposition on the lagoon side of material washed over the island during storms. The island has discontinuous dunes 10 to 15 feet high, but the system has been damaged by development. Evacuation during a storm is restricted to a single route, the Queen Isabella Causeway. Although the causeway itself has a high elevation, the approach roads on the island are low and likely to be flooded during early stages of a storm (fig. 4.5).

Seawalls on the Gulf beach have been built in front of most of the condominiums. Most of the seawalls were destroyed or severely damaged in 1980 by Hurricane Allen. A seawall in front of one single-family dwelling has failed twice (fig. 3.6), the last time in Hurricane Buelah (1967) which had a storm surge of over 7 feet in this area. In general, seawall construction on South Padre is leading to increased beach narrowing and to a less attractive natural environment. Adding to the problem is a long history of removing sand from dunes to increase the elevation of construction sites. The long-range outlook (30 to 40 years) for South Padre Island is bleak. Unless preventive measures are taken it is safe to predict that seawalls will become nearly continuous over the next few years, that beaches will narrow and almost disappear, and that eventually the remaining pocket beaches will be littered with debris from failed shoreline structures.

Brazos Santiago Pass to State Highway No. 4 (Brazos Island: figure 4.39)

This barrier segment is actually part of the Rio Grande delta that is retreating at rates of 5 to 10 feet per year. The rate of erosion decreases northward toward the jetties. Jetty construction at Brazos Santiago Pass began as early as 1881, but early attempts were abandoned and destroyed until the present jetties were completed in 1935.

Contributing to the shoreline erosion problem is depletion of the sand supply from the Rio Grande caused by damming upstream. Dune heights along this stretch are irregular and elevations are low. Hence, flooding is frequent, and the island is overwashed where dunes are low. One overwash channel is located at the former position of Boca Chica Pass, a tidal inlet that was active during the late 1800s. The storm surge recorded for Hurricane Beulah along this beach was 7.5 feet. The area is undeveloped, and access and evacuation are by four-wheel-drive vehicle.

State Highway No. 4 to mouth of the Rio Grande (figure 4.39)

This stretch of the Rio Grande delta generally has a narrow beach and very low elevations; dunes are low and discontinuous, and the area is susceptible to flooding and frequent overwash. The

110 Living with the Texas shore

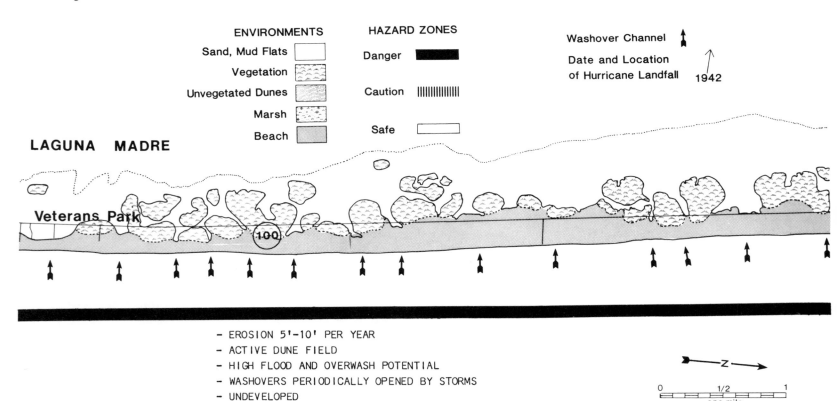

Fig. 4.38. Site analysis: South Padre Island from La Punta Larga to Veterans Park.

4. Selecting a site 111

LAGUNA MADRE

La Punta Larga

SOUTH PADRE ISLAND

1910

- EROSION 5'–10' PER YEAR
- ACTIVE DUNE FIELDS
- HIGH FLOOD AND OVERWASH POTENTIAL
- WASHOVERS PERIODICALLY OPENED BY STORMS
- UNDEVELOPED

112   Living with the Texas shore

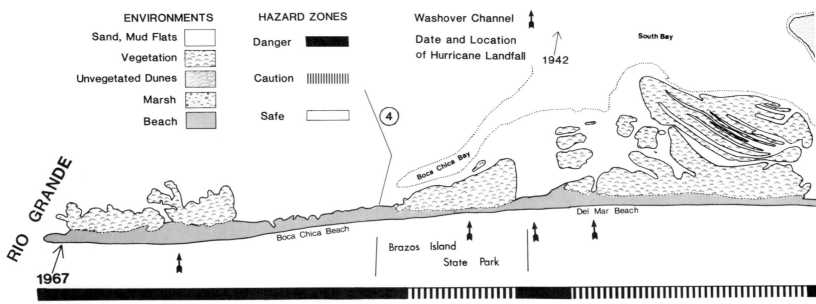

Fig. 4.39. Site analysis: South Padre Island and Boca Chica Island to the Rio Grande.

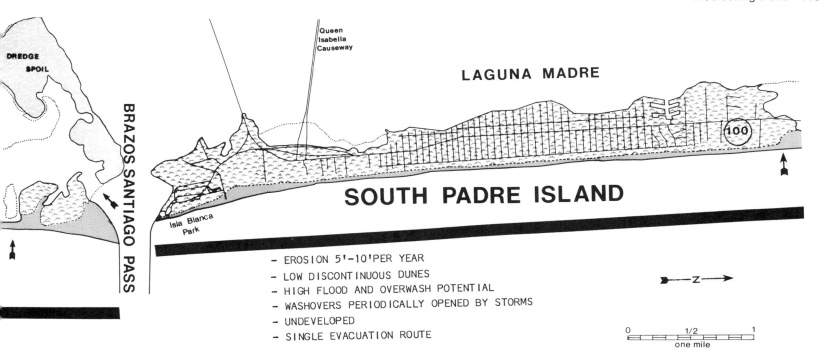

Fig. 4.40. Condominium on South Padre Island and similar structure at Panama City damaged by Hurricane Eloise in 1975.

area is undeveloped and accessible only by driving on the beach. The Gulf shoreline is eroding at 10 to 20 feet per year. These high erosion rates are partially caused by the erratic changes in position of the Rio Grande. The river mouth has migrated in a northerly direction, but flooding during major storms has caused it to return to a more southerly position.

Evacuation of the area between Brazos Santiago Pass and the Rio Grande is over a single route (State Highway No. 4) that is easily flooded because of low elevations.

5. The barrier coast of Texas, land use, and the law

Current and prospective owners of coastal property in Texas should be aware of their responsibilities under current law with respect to development and land use, and of the likelihood of later regulation. Following is a partial list of land-use regulations presently applicable to the Texas coast. The explanations we have provided are general; appendix B lists the agencies that will supply more specific information. For a more thorough description of coastal legislation and regulation in Texas, see the article by Peter Graber in the journal *Shore and Beach* (reference 90, appendix C).

The National Flood Insurance Program

One of the most significant legal pressures applied to encourage land-use planning and management in the coastal zone is the National Flood Insurance Program (NFIP). The National Flood Insurance Act of 1968 (P.L. 90-448) as amended by the Flood Disaster Protection Act of 1973 (P.L. 92-234) requires that certain conditions be met for the purchase of flood insurance. Federal assistance for building or acquiring structures—for instance VA-guaranteed mortgages, FHA mortgage insurance, loans from the Small Business Administration, or loans from the Farmers Home Administration—in flood-prone areas are available only if the community participates in the NFIP and the homeowner purchases flood insurance. Persons who do not purchase flood insurance are not eligible for most forms of federal disaster assistance in the event of a flood. Although the law is often associated with river floodplains, it also applies to barrier islands and coastal areas subject to storm-surge flooding and waves (see chapter 2, under Hurricane forces: wind, waves, and washover).

The initiative for qualifying for the program rests with the community, which must contact the Federal Emergency Management Agency (FEMA) (see appendices B and C). Once the community adopts initial land-use measures and applies for eligibility, FEMA designates the community as eligible for subsidized insurance under the Emergency Program. During this time a Flood Hazard Boundary Map that delineates flood-prone areas is in effect. Ultimately, FEMA conducts a rate study and provides the community with a Flood Insurance Rate Map and 100-year-flood elevation data. (The level that flood waters are expected to rise to only once in a hundred years is referred to as the 100-year flood level; a flood of this severity has a one percent probability of occurrence in any given year.) In addition, velocity zones (V-Zones) are designated on these maps. These zones are areas likely to be penetrated by waves on top of the storm-surge flood level. Subsidized insurance is available for all existing and new construction except for those buildings constructed in the flood-hazard area *after* publication of the Flood Insurance Rate Map. New structures, or significantly improved older structures, must meet elevation requirements to comply with local flood-plan management regulations and must pay actuarial (true risk) rates for the insurance. As the cost of the coastal flood experience goes up, rates are likely to be adjusted upward. Since October 1981, buildings placed in V-Zones must meet higher elevation requirements to accommodate potential wave height, and have been charged higher actuarial rates. General eligibility requirements vary among pole houses, mobile homes, and condominiums.

Before building or buying a home, an individual should ask certain basic questions:

Is the community I'm locating in covered by the program?

Is my building site above the 100-year flood level (plus wave height in the V-Zone)?

What are the structural requirements for my building?

What are the premiums and limits of coverage?

Most lending institutions and building inspectors will be aware of mapped flood-prone areas, but it would be wise to confirm such information with the appropriate insurance representative or program office (see appendix B, under Insurance). To obtain the best flood protection, owners should know the difference between a homeowner's policy and a flood-insurance policy. Homeowners' insurance only covers structural damage from wind or wind-driven rain whereas flood insurance covers losses due to inundation by stream flooding, storm surge, or runoff of surface water. Flood insurance is available for residential or commercial buildings. If you own the building you can insure either structure or contents, or both; if you rent the building you need only insure the contents.

The flood insurance program undoubtedly has its flaws. Some feel that requirements for insurance eligibility should be more stringent and they suggest the possibility of denying flood insurance and federal-disaster assistance to current property owners located in coastal high-hazard areas that do not meet the standards for new construction. Other problems include determining and establishing actuarial rates, funding the necessary coastal studies to define high-hazard areas, and understanding the effects of long-term federal subsidy. Perhaps the greatest obstacle to the success of the program is the uninformed individual who stands to gain the most from it. One study examined public response to flood insurance and found that many people have little awareness of the threat of floods or the cost of insurance, and view insurance as an investment with the expectation of a return rather than a means of sharing the cost of natural disaster.

Prospective homeowners should be aware of moves to alter the structure of the flood insurance program. It seems likely that in the near future a much greater share of the cost for the program will be shifted away from the taxpayer, to the property owner, causing a further rise in premiums.

Coastal Barriers Resources Act

The Coastal Barriers Resources Act (P.L. 97-348) passed by Congress in 1982 will prohibit federal financing of projects and flood insurance on undeveloped barrier islands after October 1983. After that date projects such as highway, bridge, and utility construction as well as home mortgages will be ineligible for federal funds.

Areas designated as undeveloped by the U.S. Department of Interior included about 150 miles of the Texas coast, or nearly one-fifth of the 800 miles of coast affected nationwide. In addition to the 150 miles of Texas coast now covered by the act, formerly all of Mustang Island and a 7.5-mile stretch of South Padre Island were included in the undeveloped category; these were deleted, however, upon the request of high-ranking state officials.

Waste disposal

Protecting coastal water resources is essential to safeguarding the many ways in which man uses the land. Fisheries, all forms of water recreation, and the general ecosystem depend on high-quality surface waters. Potable water supplies are also drawn mainly from surface water which must be of high quality. As noted in chapter 4, water resources in some areas are being threatened, and the existing pollution is costly to both local communities and the state.

Improper land use relative to waste disposal and inadequate planning for treatment of the increasing waste necessitate additional regulation at all levels of government.

The Federal Water Pollution Control Act Amendments of 1972 (P.L. 92-500) control any type of land use that generates—or may generate—water pollution. They also regulate dredging and filling of wetlands and water bodies through the Army Corps of Engineers (see appendix B, under Dredging, Filling, and Construction in Coastal Waterways; Sanitation; and Water Resources). The Marine Protection, Research and Sanctuaries Act of 1972 (P.L. 92-532) regulates dumping into ocean water. The Water Resources Development Act of 1974 (P.L. 93-251) also provides for comprehensive coastal-zone planning.

The Texas Department of Water Resources and the Department of Health share jurisdiction in approving sanitary-disposal systems through a permit system (see appendix B, under Sanitation). As called for in the Texas Water Code, the Department of Water Resources establishes water qualilty standards and regulates waste discharge and private sewage facilities. Responsibilities of the Department of Health include reviewing plans and specifications for sewage-disposal systems for public use, and certifying water and sewer operations.

Building codes

Most progressive communities require that new construction adhere to the provisions of a recognized building code. If you plan to build in an area that does not follow such a code, you would be wise to insist that your builder do so. Local building officials in storm areas often adopt national codes that contain building requirements for protection against high wind and water. Compiled by knowledgeable engineers, politicians, and architects, these codes regulate the design and construction of buildings and the quality of building materials. Examples of such codes are the Southern

Standard Building Code (reference 94, appendix C), used chiefly on the Southeast and Gulf coasts, and the Uniform Building Code (reference 96, appendix C), used mainly on the West Coast. Both are excellent codes. Except where local municipalities such as Galveston or Corpus Christi have established their own code, Texas builders must adhere to the Southern Standard Building Code.

It is emphasized that the purpose of these codes is to provide *minimum* standards to safeguard lives, health, and property. These codes protect you from yourself as well as from your neighbor.

Open beaches

The Open Beaches Act, passed by the Texas legislature in 1959, defines the rights of the public on the beaches bordering the Gulf of Mexico. As written, the Open Beaches Act does not apply to beaches bordering islands or peninsulas that are inaccessible to the public, such as San Jose and Matagorda Islands. A right-of-way established by the act provides access to the public beach and includes the area from mean low water up to the line of continuous vegetation. This area includes the state-owned beaches, which extend from mean low water to mean high water, and some privately owned land areas between mean high water and the vegetation line.

The law has generated several lawsuits so far, and additional litigation is anticipated as a result of movement of the vegetation line which is both a geological and legal boundary. Perhaps the most difficult problem will be to resolve the conflicts between public rights and private rights in areas where the Gulf shoreline is eroding.

Excavation of sand

This law provides for the regulation of removal of material (sand, marl, gravel, or shell) from barrier islands and peninsulas. Applications for excavation are made through the appropriate county commissioners' court. The court reviews the application and issues a permit only after determining that the operation will not create hazardous conditions that would invite hurricane washover. This act prohibits the removal of sand from state land or the public beach except as stated in the provisions.

Among other exemptions are (1) the use of sand within the property on which it occurs, and (2) construction improvements on the property where sand is excavated. This act also does not apply to any island or peninsula that is inaccessible to the public.

Sand dune protection

This legislative act recognizes the important role dunes play in storm protection and control of shoreline erosion. It also recognizes the detrimental effect of recreational vehicles and other activities that destroy dune vegetation. The bill allows the appropriate county commissioners' court to establish a dune protection line after holding public hearings. Maximum extent of the dune protection line is limited to 1,000 feet landward of the mean high-water line of the Gulf. Once the dune-protection line is established, the county commissioners' court is empowered to implement a permit system to regulate activities in the dunes.

Permits are granted by the county commissioners' court if evaluations indicate that the function of the dunes would not be weakened. Activities not covered by the bill include (1) livestock grazing, (2) oil and gas production, and (3) recreational activity other than that relating to recreational vehicles.

The bill divides the coast into three segments. The first segment is the upper coast from Sabine Pass to Aransas Pass. Activities involving dunes seaward of the dune-protection line in this segment require permits if removal of dune sand or destruction of dune vegetation is anticipated. The second segment is Aransas Pass to Mansfield Ship Channel. Dunes in this segment are protected to the extent that reduction in elevation below that shown on the Federal Flood Hazard Map is prohibited. Also in this segment, activities that would destroy the vegetation require permits and provisions for dune stabilization. In both segments, recreational vehicles are prohibited on dunes seaward of the dune-protection line.

Concerning the third segment, the bill finds "that the area bounded on the north by the Mansfield Ship Channel and extending to the southern tip of South Padre Island is an area of irregular dunes, unstable, and migratory; and that such dunes do not afford significant protection to persons and property inland from this area." Although the bill does not apply south of Mansfield Channel, the dunes on Brazos Island between Brazos Santiago Pass and the mouth of the Rio Grande were not specifically excluded. Furthermore, the act does not apply to any island or peninsula that is inaccessible to the public.

The Texas General Land Office (GLO) is charged with the responsibility of delineating critical dune areas related to the protection of state land. Nueces County, the first county to establish a dune-protection line under this bill, chose to include the entire area 1,000 feet landward from mean high water. The GLO currently reviews permit applications when requested by the county.

The Texas Coastal Public Lands Management Act

In 1973, the Coastal Public Lands Management Act (CPLMA) was passed by the Texas legislature to insure good land-use and resource development of *state-owned land*, and conservation of *state resources* to protect the quality of life for citizens of the coastal zone. It was prompted by the growing recognition of the barrier-island and adjacent bay systems as natural resources, and the increasing realization that actions in one area of an island or adjacent submerged land may affect all other resources within the system.

CPLMA is timely in that it fosters the conservation of state-owned submerged lands adjacent to barrier islands through required planning and the coordination of existing laws that control development (see appendix B, under Dredging, Filling, and Construction in Coastal Waterways; Dune Alteration; and Sanitation). Though the law is modern, it may encourage citizens of the 1980s to develop the virtues of the old-timers: prudence in building, foresight in responding to physical processes, wisdom in planning, and thrifty utilization of the natural resources at hand.

The Texas Coastal Program

Texas has been working toward obtaining approval of a coastal management program since 1973 when the governor designated the General Land Office (GLO) as lead agency in this effort. Currently the Texas Energy and Natural Resources Advisory Council (TENRAC) is responsible for the state's program. However, final approval of a state program has not been obtained and federal funds for the work were terminated in 1981.

The cooperation of various state agencies with responsibilities in the coastal zone has led to many useful planning documents that are presented in the draft Texas Coastal Program (TCP). Throughout program formulation, the goal of the TCP was not to stop or prohibit development but rather to control it. Yet it met with opposition from coastal residents. It promoted planning, which some regard as inconvenient, and a change in attitude or behavior that would interfere with the rapid exploitation of our coastal resources.

Even though the program was not implemented, we encourage coastal residents to obtain a copy of the draft Texas Coastal Program (reference 89, appendix C), which presents a more thorough treatment of the physical, social, political, and legal questions at issue along the coast.

6. Building or buying a house near the beach

In reading this book you may conclude that the authors seem to be at cross-purposes. On the one hand, we point out that building on the coast is risky. On the other hand, we provide you with a guide to evaluate the risks, and in this chapter we describe how best to buy or build a house near the beach.

This apparent contradiction is more rational than it might seem at first. For those who will heed the warning, we describe the risks of owning shorefront property. But we realize that coastal development will continue. Some individuals will always be willing to gamble with their fortunes to be near the shore. For those who elect to play this game of real estate roulette, we provide some advice on improving the odds, on reducing (not eliminating) the risks. We do not recommend, however, that you play the game!

If you want to learn more about construction near the beach, we recommend the book *Coastal Design: A Guide for Builders, Planners, and Homeowners* (Van Nostrand Reinhold Company, 1983), which gives more detail on coastal construction and supplements this Texas volume. In addition, the Federal Emergency Management Agency's *Design and Construction Manual for Residential Buildings in Coastal High Hazard Areas* is an excellent guide to coastal construction and additional reference material. See references 92 and 93, appendix C.

Coastal realty versus coastal reality

Coastal property is not the same as inland property. Do not approach it as if you were buying a lot in a developed woodland in western New York or a wheat field in Kansas. The previous chapters illustrate that open-ocean coasts, especially barrier islands, are composed of variable environments and are subjected to nature's most powerful and persistent forces. The reality of the coast is its dynamic character. Property lines are an artificial grid superimposed on this dynamism. If you choose to place yourself or others in this zone, prudence is in order.

A quick glance at the architecture of the structures on our coast provides convincing evidence that the reality of coastal processes was rarely considered in their construction. Apparently the sea view and aesthetics were the primary considerations. Except for meeting minimal building code requirements, no further thought seems to have been given to the safety of many of these buildings. The failure to follow a few basic architectural guidelines that recognize this reality will have disastrous results in the next major storm.

Life's important decisions are based on an evaluation of the facts. Few of us buy goods, choose a career, or take legal, financial, or medical actions without first evaluating the facts and seeking advice. In the case of coastal property, two general aspects should be evaluated: site safety, and the integrity of the structure relative to the forces to which it will be subjected.

A guide to evaluating the site(s) of your interest on the Texas open-ocean shoreline is presented in chapter 4, along with hazard evaluation maps. The remainder of this chapter focuses on the structure itself, whether cottage or condominium.

The structure: concept of balanced risk

A certain chance of failure for any structure exists within the constraints of economy and environment. The objective of building design is to create a structure that is both economically feasible and functionally reliable. A house must be affordable and have a reasonable life expectancy free of being damaged, destroyed, or wearing out. In order to obtain such a house, a balance must be achieved among financial, structural, environmental, and other special conditions. Most of these conditions are heightened on the coast—property values are higher, there is a greater desire for aesthetics, the environment is more sensitive, the likelihood of storms is increased, etc.

The individual who builds or buys a home in an exposed area should fully comprehend the risks involved and the chance of harm to home or family. The risks should then be weighed against the benefits to be derived from the residence. Similarly, the developer who is putting up a motel should weigh the possibility of destruction and death during a hurricane versus the money and other advantages to be gained from such a building. Then and only then should construction proceed. For both the homeowner and the developer, proper construction and location reduce the risks involved.

The concept of balanced risk should take into account the following fundamental considerations:

1. Construction must be economically feasible.
2. Because construction must be economically feasible, ultimate and total safety is not obtainable for most homeowners on the coast.
3. A coastal structure, exposed to high winds, waves, or flooding, should be stronger than a structure built inland.
4. A building with a planned long life, such as a year-round residence, should be stronger than a building with a planned short life, such as a mobile home.
5. A building with a high occupancy, such as an apartment building, should be safer than a building with low occupancy, such as a single-family dwelling.
6. A building that houses elderly or sick people should be safer than a building housing able-bodied people.

Structures can be designed and built to resist all but the largest storms and still be within reasonable economic limits.

Structural engineering is the designing and construction of buildings to withstand the forces of nature. It is based on a knowledge of the forces to which the structures will be subjected and an understanding of the strength of building materials. The effectiveness of structural engineering design was reflected in the aftermath of Cyclone Tracy which struck Darwin, Australia, in 1974: 70 percent of housing that was not based on structural engineering principles was destroyed and 20 percent was seriously damaged—only 10 percent of the housing weathered the storm. In contrast, over 70 percent of the structurally engineered, large commercial, government, and industrial buildings came through with little or no damage, and less than 5 percent of such structures were destroyed. Because housing accounts for more than half of the capital cost of the buildings in Queensland, state government established a building code that requires standardized structural engineering for houses in hurricane-prone areas. This improvement has been achieved with little increase in construction and design costs.

Coastal forces: design requirements

Although nontropical storms can also be devastating, hurricanes produce the most destructive forces to be reckoned with on the coast (fig. 6.1).

Hurricane winds

Hurricane winds can be evaluated in terms of the pressure they exert. A 100-mph wind exerts a pressure or force of about 40 pounds per square foot on a flat surface. The pressure varies with the square of the velocity. For example, a wind of 190-mph velocity exerts a force of 144 pounds per square foot. This force is modified by several factors which must be considered in designing a building. For instance, the effect on a round surface, such as that of a sphere or cylinder, is less than the effect on a flat surface. Also, winds increase with height above ground, so a tall structure is subject to greater pressure than a low structure.

A house or building designed for inland areas is built primarily to resist vertical loads. It is assumed that the foundation and framing must support the load of the walls, floor, and roof, and relatively insignificant horizontal wind forces.

A well-built house in a hurricane-prone area, however, must be constructed to withstand a variety of strong wind forces that may come from any direction. Although many people think that wind damage is caused by uniform horizontal pressures (lateral loads), most damage, in fact, is caused by uplift (vertical), suctional (pressure-differential), and torsional (twisting) forces. High horizontal pressure on the windward side is accompanied by suction on the leeward side. The roof is subject to downward pressure and, more importantly, to uplift. Often a roof is sucked up by the uplift drag of the wind. Usually the failure of houses is in the devices that tie the parts of the structure together. All structural members (beams, rafters, columns) should be fastened together on the assumption that about 25 percent of the vertical load on the member may be a force coming from any direction (sideways or upwards). Such structural integrity is also important if it is likely that the building may be moved to avoid destruction by shoreline retreat.

A fanciful way of understanding structural integrity is that you should be able to pick up a house (after having removed its

124 Living with the Texas shore

WIND

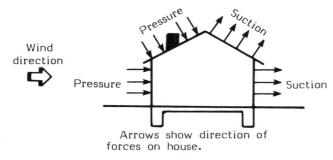

Arrows show direction of forces on house.

DROP IN BAROMETRIC PRESSURE

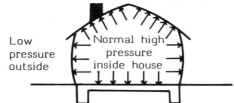

The passing eye of the storm creates different pressure inside and out, and high pressure inside attempts to burst house open.

WAVES

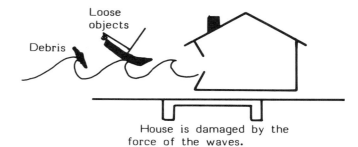

House is damaged by the force of the waves.

HIGH WATER

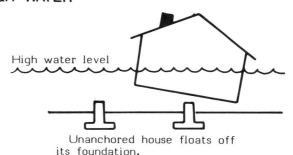

Unanchored house floats off its foundation.

Fig. 6.1. Forces to be reckoned with at the seashore.

furniture of course), turn it upside down, and shake it without it falling apart.

Storm surge

Storm surge is a rise in sea level above the normal water level during a storm. During hurricanes the coastal zone is inundated by storm surge and accompanying storm waves, and these cause most property damage and loss of life. (For more on storm surge see chapter 2.)

Often the pressure of the wind backs water into streams or estuaries already swollen from the exceptional rainfall brought on by the hurricane. Water is piled into the bays between islands and the mainland by the offshore storm. In some cases the direction of flooding may be from the bay side of the island. This flooding is particularly dangerous when the wind pressure keeps the tide from running out of inlets, so that the next normal high tide pushes the accumulated waters back and higher still.

People who have cleaned the mud and contents out of a house subjected to flooding retain vivid memories of its effects. Flooding can cause an unanchored house to float off its foundation and come to rest against another house, severely damaging both. Even if the house itself is left structurally intact, flooding may destroy its contents.

Proper coastal development takes into account the expected level and frequency of storm surge for the area. In general, building standards require that the first habitable level of a dwelling be above the 100-year flood level. At this level, a building has a one-percent probability of being flooded in any given year.

Hurricane waves

Hurricane waves can cause severe damage not only in forcing water onshore to flood buildings but also in throwing boats, barges, piers, houses, and other floating debris inland against standing structures. In addition, waves can destroy coastal structures by scouring away the underlying sand, causing them to collapse. It is possible to design buildings to survive crashing storm surf. Many lighthouses, for example, have survived this. But in the balanced-risk equation, it usually isn't economically feasible to build ordinary cottages to resist the more powerful of such forces. On the other hand, cottages can be made considerably more storm-worthy by following the suggestions in the upcoming sections.

The force of a wave may be understood when one considers that a cubic yard of water weighs over three-fourths of a ton; hence, a breaking wave moving shoreward at a speed of several tens of miles per hour can be one of the most destructive elements of a hurricane.

Barometric pressure change

Changes in barometric pressure may also be a minor contributor to structural failure. If a house is sealed at a normal barometric pressure of 30 inches of mercury, and the external pressure suddenly drops to 26.61 inches of mercury (as it did in Hurricane

Camille in Mississippi in 1969), the pressure exerted within the house would be 245 pounds per square foot. An ordinary house would explode if it were leakproof. In tornadoes, where there is a severe pressure differential, many houses do just that. In hurricanes the problem is much less severe. Fortunately, most houses leak, but they must leak fast enough to prevent damage. Given the more destructive forces of hurricane wind and waves, pressure differential may be of minor concern. Venting the underside of the roof at the eaves is a common means of equalizing internal and external pressure.

Figure 6.2 illustrates some of the actions that a homeowner can take to deal with the forces just described.

House selection

Some types of houses are better than others at the shore, and an awareness of the differences will help you make a better selection, whether you are building a new house or buying an existing one.

Worst of all are unreinforced masonry houses—whether they be brick, concrete block, hollow clay-tile, or brick veneer—because they cannot withstand the lateral forces of wind and wave and the settling of the foundation.

Adequate and extraordinary reinforcing in coastal regions will alleviate the inherent weakness of unit masonry, if done properly. Reinforced concrete and steel frames are excellent but are rarely used in the construction of small residential structures.

It is hard to beat a wood-frame house that is properly braced and anchored and has well-connected members. The well-built wood house will often hold together as a unit, even if moved off its foundations, when other types disintegrate. Although all of the structural types noted above are found in the coastal zone, newer structures tend to be of the elevated wood-frame type.

Keeping dry: pole or "stilt" houses

In coastal regions subject to flooding by waves or storm surge, the best and most common method of minimizing damage is to raise the lowest floor of a residence above the expected water level. Also, the first habitable floor of a home must be above the 100-year storm-surge level (plus calculated wave height) to qualify for federal flood insurance. As a result, most modern flood-zone structures are constructed on piling, well anchored in the subsoil. Elevating the structure by building a mound is not suited to the coastal zone because mounded soil is easily eroded.

Current building-design criteria for pole-house construction under the flood insurance program are outlined in the book *Elevated Residential Structures* (reference 113, appendix C). Regardless of insurance, pole-type construction with deep embedment of the poles is best in areas where waves and storm surge will erode foundation material. Materials used in pole construction include the following:

Piles. Piles are long, slender columns of wood, steel, or concrete driven into the earth to a sufficient depth to support the vertical

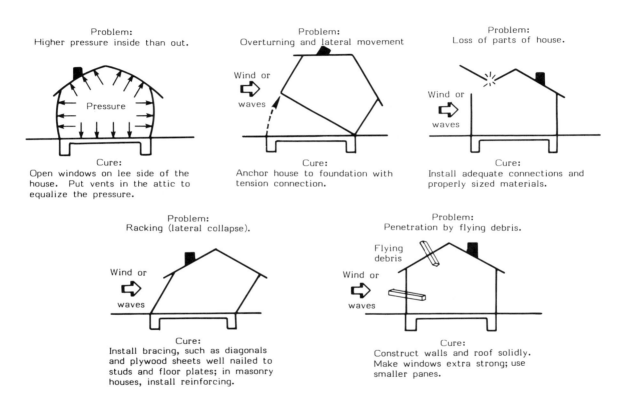

Fig. 6.2. Modes of failure and how to deal with them. Modified from: U.S. Civil Defense Preparedness Agency Publication TR-83.

load of the house and to withstand the horizontal forces of flowing water, wind, and water-borne debris. Pile construction is especially suitable in areas where scouring—soil "washing out" from under the foundation of a house—is a problem.

Posts. Posts are usually of wood; if of steel, they are called columns. Unlike piles, they are not driven into the ground but, rather, are placed in a pre-dug hole at the bottom of which may be a concrete pad (fig. 6.3). Posts may be held in place by backfilling and tamping earth or by pouring concrete into the hole after the post is in place. Posts are more readily aligned than driven piles and are, therefore, better to use if poles must extend to the roof. In general, treated wood is the cheapest and most common material for both posts and piles.

Piers. Piers are vertical supports, thicker than piles or posts, usually made of reinforced concrete or reinforced masonry (concrete blocks or bricks). They are set on footings and extend to the underside of the floor frame.

Pole construction can be of two types. The poles can be cut off at the first-floor level to support the platform that serves as the dwelling floor. In this case, piles, posts, or piers can be used. Or they can be extended to the roof and rigidly tied into both the floor and the roof. In this way, they become major framing members for the structure and provide better anchorage to the house as a whole (figs. 6.4 and 6.5). A combination of full- and floor-height poles is

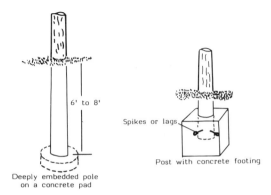

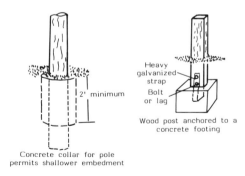

Fig. 6.3. Shallow and deep supports for poles and posts. Source: Southern Pine Association.

Fig. 6.4. Extending poles to the roof, as shown in this photograph, instead of the usual method of cutting them off at the first floor, greatly strengthens a beach cottage. Photograph by Orrin H. Pilkey, Jr.

used in some cases, with the shorter poles restricted to supporting the floor inside the house (fig. 6.6).

Where the foundation material can be eroded by waves or winds, the poles should be deeply embedded and solidly anchored either by driving piles or by drilling deep holes for posts and putting in a concrete pad at the bottom of each hole. Where the embedment is shallow, a concrete collar around the poles improves anchorage (fig 6.3). The choice depends on the soil conditions. Piles are more difficult than posts to align to match the house frame; posts can be positioned in the holes before backfilling. Inadequate piling depths, improper piling-to-floor connections, and inadequate pile bracing all contribute to structural failure when storm waves liquefy and erode sand support.

When post holes are dug, rather than pilings driven, the posts should extend 6 to 8 feet into the ground to provide anchorage. The lower end of the post should rest on a concrete pad, spreading the load to the soil over a greater area to prevent settlement. Where the soil is sandy or the embedment less than, say, 6 feet, it is best to tie the post down to the footing with straps or other anchoring devices. Driven piles should extend to a depth of 8 feet or more.

The floor and the roof should be securely connected to the poles with bolts or other fasteners. When the floor rests on poles that do not extend to the roof, attachment is even more critical. A system of metal straps is often used. Unfortunately, it is very common for builders simply to attach the floor joists to a notched pole by one or two bolts. Hurricanes have proven this method insufficient. During the next hurricane on the barrier-island coast, many houses will be destroyed because of inadequate attachment.

Local building codes may specify the size, quality, and spacing of the piles, ties, and bracing, as well as the methods of fastening the structure to them. Building codes often are minimal require-

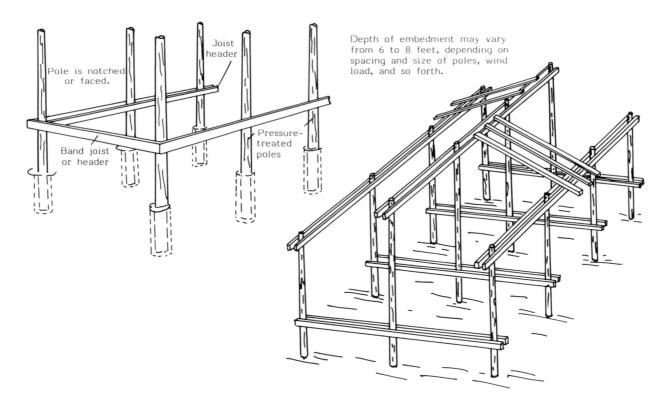

Fig. 6.5. Framing system for an elevated house. Source: Southern Pine Association.

6. Building or buying a house

ments, however, and building inspectors are usually amenable to allowing designs that are equally or more effective.

The space under an elevated house, whether pole-type or otherwise, must be kept free of obstructions in order to minimize the impact of waves and floating debris. If the space is enclosed, the enclosing walls should be designed so that they can break away or fall under flood loads but also remain attached to the house or be heavy enough to sink; thus, the walls cannot float away and add to the water-borne debris problem. Alternative ways of avoiding this problem are designing walls that can be swung up out of the path of the floodwaters, or building them with louvers that allow the water to pass through. The louvered wall is subject to damage from floating debris. The convenience of closing in the ground floor for a garage or extra bedroom may be costly because it may violate insurance requirements and actually contribute to the loss of the house in a hurricane.

An existing house: what to look for, where to improve

If instead of building a new house, you are selecting a house already built in an area subject to flooding and high winds, consider the following factors: (1) where the house is located; (2) how well the house is built; and (3) how the house can be improved.

Geographic location

Evaluate the site of an existing house using the same principles given earlier for the evaluation of a possible site to build a new

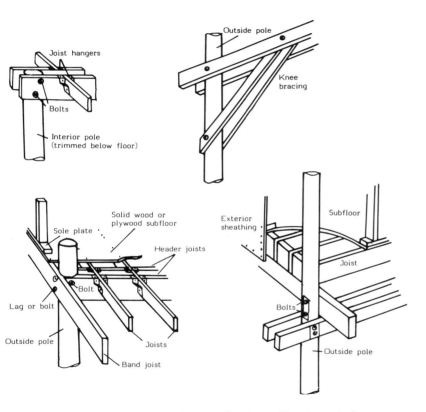

Fig. 6.6. Tying floors to poles. Source: Southern Pine Association.

house. House elevation, frequency of high water, escape route, and how well the lot drains should be emphasized, but you should go through the complete site-safety checklist given in chapter 4.

You can modify the house after you have purchased it but you can't prevent hurricanes or other storms. The first step is to stop and consider: Do the pleasures and benefits of this location balance the risks and disadvantages? If not, look elsewhere for a home; if so, then evaluate the house itself.

How well built is the house?

In general, the principles used to evaluate an existing house are the same as those used in building a new one (see references 92 to 115, appendix C).

Before you thoroughly inspect the house in which you are interested, look closely at the adjacent homes. If poorly built, they may float over against your house and damage it in a flood. You may even want to consider the type of people you will have as neighbors: Will they "clear the decks" in preparation for a storm or will they leave items in the yard to become windborne missiles? The house itself should be inspected for the following:

The house should be well anchored to the ground. If it is simply resting on blocks, rising water may cause it to float off its foundation and come to rest against your neighbor's house or out in the middle of the street. If well built and well braced internally, it may be possible to move the house back to its proper location, but chances are great that the house will be too damaged to be habitable.

If the house is on piles, posts, or poles, check to see if the floor beams are adequately bolted to them. If it rests on piers, crawl under the house if space permits to see if the floor beams are securely connected to the foundation. If the floor system rests unanchored on piers, do not buy the house.

It is difficult to discern whether a house built on a concrete slab is properly bolted to the slab because the inside and outside walls hide the bolts. If you can locate the builder, ask if such bolting was done. Better yet, if you can get assurance that construction of the house complied with the provisions of a building code serving the needs of that particular region, you can be reasonably sure that all parts of the house are well anchored: the foundation to the ground, the floor to the foundation, the walls to the floor, and the roof to the walls (figs. 6.7, 6.8, and 6.9).

Be aware that many builders, carpenters, and building inspectors who are accustomed to traditional construction are apt to regard metal connectors, collar beams, and other such devices as newfangled and unnecessary. If consulted, they may assure you that a house is as solid as a rock, when in fact, it is far from it. Nevertheless, it is wise to consult the builder or knowledgeable neighbors when possible.

The roof should be well anchored to the walls. This will prevent uplifting and separation from the walls. Visit the attic to see if

6. Building or buying a house   133

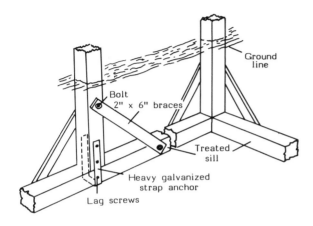

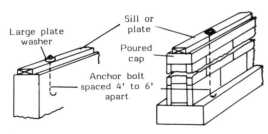

Fig. 6.7. Foundation anchorage. Top: anchored sill for shallow embedment. Bottom: anchoring sill or plate to foundation. Source of bottom drawing: *Houses Can Resist Hurricanes*, U.S. Forest Service Research Paper FPL 33.

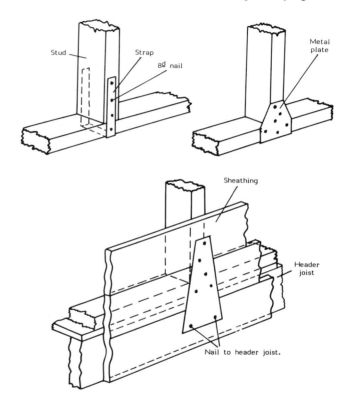

Fig. 6.8. Stud-to-floor, plate-to-floor framing methods. Source: *Houses Can Resist Hurricanes*, U.S. Forest Service Research Paper FPL 33.

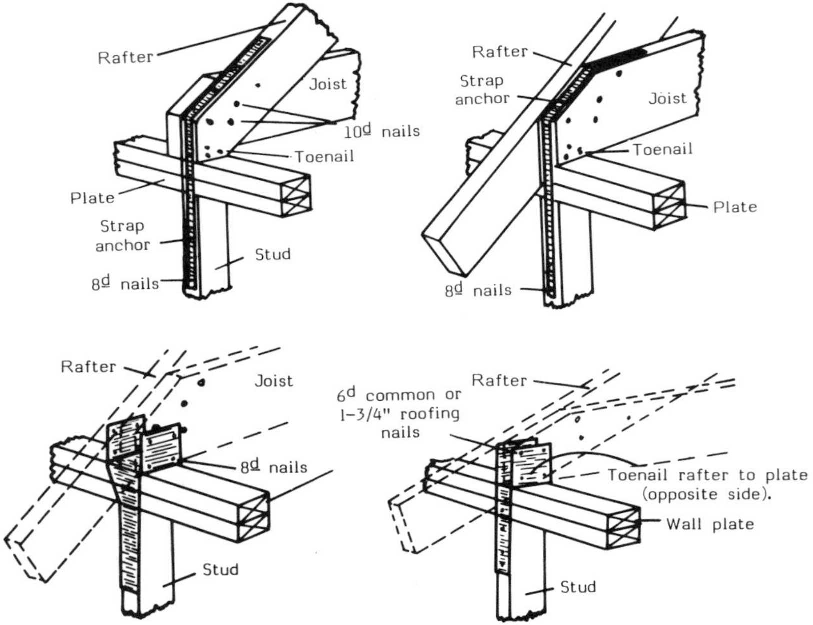

Fig. 6.9. Roof-to-wall connections. The top drawings show metal strap connectors: left, rafter to stud; right, joist to stud. The bottom left drawing shows a double-member metal-plate connector—in this case with the joist to the right of the rafter. The bottom right drawing shows a single-member metal-plate connector. Source: *Houses Can Resist Hurricanes*, U.S. Forest Service Research Paper FPL 33.

such anchoring exists. Simple toe-nailing (nailing at an angle) is not adequate; metal fasteners are needed. Depending on the type of construction and the amount of insulation laid on the floor of the attic, these may or may not be easy to see. If roof trusses or braced rafters were used, it should be easy to see whether the various members, such as the diagonals, are well fastened together. Again, simple toe-nailing will not suffice. Some builders, unfortunately, nail parts of a roof truss just enough to hold it together to get it in place. A collar beam or gusset at the peak of the roof (fig. 6.10) provides some assurance of good construction.

Quality roofing material should be well anchored to the sheathing. A poor roof covering will be destroyed by hurricane-force winds, allowing rain to enter the house and damage ceilings, walls, and the contents of the house. Galvanized nails (two per shingle) should be used to connect wood shingles and shakes to wood sheathing (fig. 6.10). Threaded nails should be used for plywood sheathing. For roof slopes that rise 1 foot for every 3 feet or more of horizontal distance, exposure of the shingle should be about one-fourth of its length (4 inches for a 16-inch shingle). If shakes (thicker and longer than shingles) are used, less than one-third of their length should be exposed.

In hurricane areas, asphalt shingles should be exposed somewhat less than usual. A mastic or seal-tab type or an interlocking shingle of heavy grade should be used. A roof underlay of asphalt-saturated felt and galvanized roofing nails or approved staples (six for each three-tab strip) should be used.

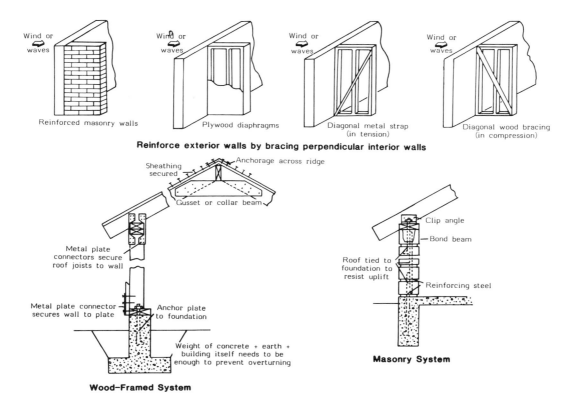

Fig. 6.10. Where to strengthen a house. Modified from: U.S Civil Defense Preparedness Agency Publication TR-83.

The fundamental rule to remember in framing is that all structural elements should be fastened together and anchored to the ground in such a manner as to resist all forces, regardless of which direction these forces may come from. This prevents overturning, floating off, racking, or disintegration.

The shape of the house is important. A hip roof, which slopes in four directions, is better able to resist high winds than a gable roof, which slopes in two directions. This was found to be true in Hurricane Camille in 1969 in Mississippi and, later, in Cyclone Tracy, which devastated Darwin, Australia, in December 1974. The reason is twofold: the hip roof offers a smaller shape for the wind to blow against, and its structure is such that it is better braced in all directions.

Note also the horizontal cross section of the house (the shape of the house as viewed from above). The pressure exerted by a wind on a round or elliptical shape is about 60 percent of that exerted on the common square or rectangular shape; the pressure exerted on a hexagonal or octagonal cross section is about 80 percent of that exerted on a square or rectangular cross section.

The design of a house or building in a coastal area should minimize structural discontinuities and irregularities. A house should have a minimum of nooks and crannies and offsets on the exterior, because damage to a structure tends to concentrate at these points. Some of the newer beach cottages along the American coast are of a highly angular design with such nooks and crannies. Award-winning architecture will be a storm loser if the design has not incorporated the technology for maximizing structural integrity with respect to storm forces. When irregularities are absent, the house reacts to storm winds as a complete unit.

Brick, concrete-block, and masonry-wall houses should be adequately reinforced. This reinforcement is hidden from view. Building codes applicable to high-wind areas often specify the type of mortar, reinforcing, and anchoring to be used in construction. If you can get assurance that the house was built in compliance with a building code designed for such an area, consider buying it. At all costs, avoid unreinforced masonry houses.

A poured-concrete bond-beam at the top of the wall just under the roof is one indication that the house is well built (fig. 6.11). Most bond beams are formed by putting in reinforcing and pouring concrete in U-shaped concrete blocks. From the outside, however, you can't distinguish these U-shaped blocks from ordinary ones and therefore can't be certain that a bond beam exists; the vertical reinforcing should penetrate the bond beam.

Some architects and builders use a stacked bond (one block directly above another), rather than overlapped or staggered blocks, because they believe it looks better. The stacked bond is definitely weaker than the latter. Unless you have proof that the walls are adequately reinforced to overcome this lack of strength, you should avoid this type of construction.

In past hurricanes, the brick veneer of many houses has separated from the wood frame, even when the houses remained standing. Asbestos-type outer-wall panels used on many houses in

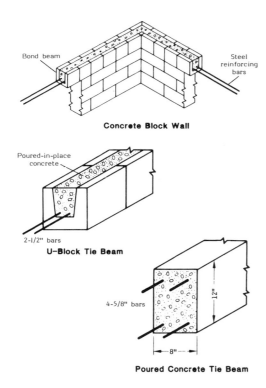

Fig. 6.11. Reinforced tie beam (bond beam) for concrete block walls —to be used at each floor level and at roof level around the perimeter of the exterior walls.

Darwin, Australia, were found to be brittle and they broke up under the impact of windborne debris in Cyclone Tracy. Both types of construction should be avoided along the coast.

Consult a good architect or structural engineer for advice if you are in doubt about any aspects of a house. A few dollars spent for wise counsel may save you from later financial grief.

To summarize, a beach house should have: (1) roof tied to walls, walls tied to foundation, and foundation anchored to the earth (the connections are potentially the weakest link in the structural system); (2) a shape that resists storm forces; (3) floors high enough (sufficient elevation) to be above most storm waters (usually the 100-year flood level plus 3 to 8 feet); (4) piles that are of sufficient depth or embedded in concrete to anchor the structure and to withstand erosion; and (5) piling that is well braced.

What can be done to improve an existing house?

If you presently own a house or are contemplating buying one in a hurricane-prone area, you will want to know how to improve occupant protection in the house. If so, you should obtain the excellent publication, *Wind-Resistant Design Concepts for Residences*, by Delbart B. Ward (reference 103, appendix C). Of particular interest are the sections on building a refuge shelter module within a residence. Also noteworthy are two supplements to this publication (reference 104, appendix C) which deal with buildings larger than single-family residences and which may be of interest to the general public, especially residents in urban areas. These provide a means of checking whether the responsible authorities are doing their jobs to protect schools, office buildings, and apartments. Several other pertinent references are listed in appendix C.

Suppose your house is resting on blocks but not fastened to them and, thus, is not adequately anchored to the ground. Can anything be done? One solution is to treat the house like a mobile home by screwing ground anchors into the ground to a depth of 4 feet or more and fastening them to the underside of the floor systems. See figures 6.12 and 6.13 for illustrations of how ground anchors can be used.

Calculations to determine the needed number of ground anchors will differ between a house and a mobile home, because each is affected differently by the forces of wind and water. Note that recent practice is to put these commercial steel-rod anchors in at an angle in order to align them better with the direction of the pull. If a vertical anchor is used, the top 18 inches or so should be encased in a concrete cylinder about 12 inches in diameter. This prevents the top of the anchor rod from bending or slicing through the wet soil from the horizontal component of the pull.

Diagonal struts, either timber or pipe, may also be used to anchor a house that rests on blocks. This is done by fastening the upper ends of the struts to the floor system, and the lower ends to individual concrete footings substantially below the surface of the ground. These struts must be able to take both uplift (tension) and compression, and should be tied into the concrete footing with anchoring devices such as straps or spikes.

If the house has a porch with exposed columns or posts, it should be possible to install tiedown anchors on their tops and bottoms. Steel straps should suffice in most cases.

When accessible, roof rafters and trusses should be anchored to the wall system. Usually the roof trusses or braced rafters are sufficiently exposed to make it possible to strengthen joints (where two or more members meet) with collar beams or gussets, particularly at the peak of the roof (fig. 6.10).

A competent carpenter, architect, or structural engineer can review the house with you and help you decide what modifications are most practical and effective. Do not be misled by someone who is resistant to new ideas. One builder told a homeowner, "You don't want all those newfangled straps and anchoring devices. If you use them, the whole house will blow away, but if you build in the usual manner [with members lightly connected], you may lose only part of it."

In fact, the very purpose of the straps is to prevent any or all of the house from blowing away. The Standard Building Code (previously known as the Southern Standard Building Code and still frequently referred to by that name) says, "Lateral support securely anchored to all walls provides the best and only sound structural stability against horizontal thrusts, such as winds of exceptional velocity" (see reference 94, appendix C). And the cost of connecting all elements securely adds very little to the cost of the frame of the dwelling, usually under 10 percent, and a very much smaller percentage to the total cost of the house.

If the house has an overhanging eave and there are no openings on its underside, it may be feasible to cut openings and screen them. These openings keep the attic cooler (a plus in the summer) and help to equalize the pressure inside and outside of the house during a storm with a low-pressure center.

Another way a house can be improved is to modify one room so that it can be used as an emergency refuge in case you are trapped in a major storm. (This is *not* an alternative to evacuation prior to a hurricane.) Examine the house and select the best room to stay in during a storm. A small, windowless room such as a bathroom, utility room, den, or storage space is usually stronger than a room with windows. A sturdy inner room, with more than one wall between it and the outside, is safest. The fewer doors, the better; an adjoining wall or baffle-wall shielding the door adds to the protection.

Consider bracing or strengthening the interior walls. Such reinforcement may require removing the surface covering and installing plywood sheathing or strap bracing. Where wall studs are exposed, bracing straps offer a simple way to achieve needed reinforcement against the wind. These straps are commercially produced and are made of 16-gauge galvanized metal with prepunched holes for nailing. These should be secured to studs and wall plates as nail holes permit (fig. 6.10). Bear in mind that they are good only for tension.

If, after reading this, you agree that something should be done to your house, do it now. Do not put it off until the next hurricane or other storm hits you!

Mobile homes: limiting their mobility

Because of their light weight and flat sides, mobile homes are vulnerable to the high winds of hurricanes, tornadoes, and severe storms. Such winds can overturn unanchored mobile homes or smash them into neighboring homes and property. Nearly six million Americans live in mobile homes today and the number is growing. Twenty to thirty percent of single-family housing production in the United States consists of mobile homes. High winds damage or destroy nearly five thousand of these homes every year, and the number will surely rise unless protective measures are taken. As one man whose mobile home was overturned in Hurricane Frederic (1979) so aptly put it, "People who live in flimsy houses shouldn't have hurricanes."

Several lessons can be learned from past experiences in storms. First, mobile homes should be located properly. After Hurricane Camille (1969), it was observed that where mobile-home parks were surrounded by woods and where the units were close together, damage was minimized, caused mainly by falling trees. In unprotected areas, however, many mobile homes were overturned and often destroyed from the force of the wind. The protection afforded by trees is greater than the possible damage from falling limbs. Two or more rows of trees are better than a single row, and trees 30 feet or more in height give better protection than shorter ones. If possible, position the mobile home so that the narrow side faces the prevailing winds.

Locating a mobile home in a hilltop park will greatly increase

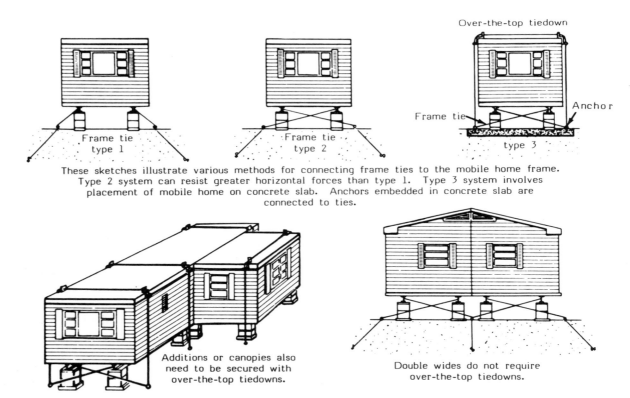

Fig. 6.12. Tiedowns for mobile homes. Source: U.S. Civil Defense Preparedness Agency Publication TR-75.

6. Building or buying a house 141

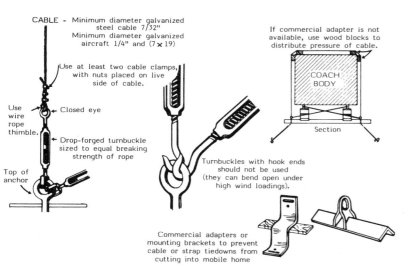

Fig. 6.13. Hardware for mobile-home tiedowns. Modified from: U.S. Civil Defense Preparedness Agency Publication TR-75.

its vulnerability to the wind. A lower site screened by trees is safer from the wind, but it should be above storm-surge flood levels. A location that is too low, obviously, increases the likelihood of flooding. There are fewer safe locations for mobile homes than for stilt houses.

A second lesson taught by past experience is that the mobile home must be tied down or anchored to the ground so that it will not overturn in high winds (figs. 6.12 and 6.13 and table 6.1). Simple prudence dictates the use of tiedowns, and in many communities ordinances require it. Many insurance companies, moreover, will not insure mobile homes unless they are adequately anchored with tiedowns.

A mobile home may be tied down with cable, rope, or built-in straps or it may be rigidly attached to the ground by connecting it to a simple wood-post foundation system. An alert owner of a mobile-home park can provide permanent concrete anchors or piers to which hold-down ties may be fastened. In general, an entire tiedown system costs only a nominal amount.

A mobile home should be properly anchored with both ties to the frame and over-the-top straps; otherwise it may be damaged by sliding, overturning, or tossing. The most common cause of major damage is the tearing away of most or all of the roof. When this happens the walls are no longer adequately supported at the top, and are more prone to collapse. Total destruction of a mobile home is more likely if the roof blows off, especially if the roof blows off first and then the home overturns. The necessity for anchoring cannot be overemphasized: there should be over-the-top tiedowns to resist overturning, and frame ties to resist sliding off the piers. This applies to single mobile homes up to 14 feet in width. "Double wides" do not require over-the-top ties but they do require frame ties.

Mobile-home owners should be sure to obtain a copy of the booklet *Protecting Mobile Homes from High Winds* (reference 114, appendix C), which treats the subject in more detail. The

Table 6.1: Tiedown anchorage requirements

Wind velocity m.p.h.	10- and 12-ft.-wide mobile homes				12- and 14-ft.-wide mobile homes, 60 to 70 ft. long	
	30 to 50 ft. long		50 to 60 ft. long			
	No. of frame ties	No. of over-the-top ties	No. of frame ties	No. of over-the-top ties	No. of frame ties	No. of over-the-top ties
70	3	2	4	2	4	2
80	4	3	5	3	5	3
90	5	4	6	4	7	4
100	6	5	7	5	8	6
110	7	6	9	6	10	7

booklet lists specific steps that one should take on receiving a hurricane warning and suggests a type of community shelter for a mobile-home park. It also includes a map of the United States with lines that indicate areas subject to the strongest sustained winds.

High-rise buildings: the urban shore

A high-rise building on the beach is generally designed by an architect and a structural engineer who are presumably well qualified and aware of the requirements for building on the shoreline. Tenants of such a building, however, should not assume that it is therefore invulnerable. Many people living in apartment buildings of two or three stories were killed when the buildings were destroyed by Hurricane Camille in Mississippi in 1969. Storms have smashed five-story buildings in Delaware. Larger high rises have yet to be thoroughly tested by a major hurricane.

The first aspect of high-rise construction that a prospective apartment dweller or condo owner must consider is the type of piling used. High rises near the beach should be built so that even if the foundation is severely undercut during a storm the building will remain standing. It is well known in construction circles that shortcuts are sometimes taken by the less scrupulous builders, and piling is not driven deeply enough. Just as important as driving the piling deep enough to resist scouring and to support the loads they must carry is the need to fasten piles securely to the structure above them which they support. The connections must resist horizontal loads from wind and wave during a storm and also uplift from the same sources. It is a joint responsibility of builders and building inspectors to make sure the job is done right. In 1975 Hurricane Eloise exposed the foundation of a just-under-construction high rise in Florida, revealing that some of the piling was not attached to the building. This happened in Panama City, Florida, but such problems probably exist everywhere that high rises crowd the beach.

Despite the assurances that come with an engineered structure, life in a high-rise building holds definite drawbacks that prospective tenants should take into consideration. The negative conditions that must be evaluated stem from high wind, high water, and poor foundations.

Pressure from the wind is greater near the shore than it is inland, and it increases with height. If you are living inland in a two-story house and plan to move to the eleventh floor of a high rise on the shore, you should expect five times more wind pressure than you are accustomed to. This can be a great and possibly devastating surprise.

The high wind-pressure actually can cause unpleasant motion of the building. It is worthwhile to check with current residents of a high rise to find out if it has undesirable motion characteristics; some have claimed that the swaying is great enough to cause motion sickness. More seriously, high winds can break windows and damage other property, and of course they can hurt people. Tenants of severely damaged buildings will have to relocate until repairs are made.

Those who are interested in researching the subject further—even the knowledgeable engineer or architect who is engaged to design a structure near the shore—should obtain a copy of *Structural Failures: Modes, Causes, Responsibilities* (reference 102, appendix C). Of particular importance is the chapter entitled, "Failure of Structures Due to Extreme Winds." This chapter analyzes wind damage to engineered high-rise buildings from the storms at Lubbock and Corpus Christi, Texas, in 1970.

Another occurrence that affects a multifamily, high-rise building more seriously than a low-occupancy structure is a power failure or blackout. Such an occurrence is more likely along the coast than inland because of the more severe weather conditions associated with coastal storms. A power failure can cause great distress. People can be caught between floors in an elevator. New York City had that experience not too long ago. Think of the mental and physical distress after several hours of confinement. And compound this with the roaring winds of a hurricane whipping around the building, sounding like a freight train. In this age of electricity, it is easy to imagine many of the inconveniences that can be caused by a power failure in a multistory building.

Fire is an extra hazard in a high-rise building. Even recently constructed buildings seem to have difficulties. The television pictures of a woman leaping from the window of a burning building in New Orleans rather than be incinerated in the blaze are a horrible reminder from recent history. The number of hotel fires of the last few years demonstrates the problems. Fire Department equipment reaches only so high. And many areas along the coast are too sparsely populated to afford high-reaching equipment.

Fire and smoke travel along ventilation ducts, elevator shafts, corridors, and similar passages. The situation *can be* corrected and the building made safer, especially if it is new. Sprinkler systems should be operated by gravity water systems rather than by powered pumps (because of possible power failure). Gravity systems use water from tanks higher up in the building. Battery-operated emergency lights that come on only when the other lights fail, better fire walls and automatic sealing doors, pressurized stairwells, and emergency-operated elevators in pressurized shafts will all contribute to greater safety. Unfortunately all of these improvements cost money, and that is why they are often omitted unless required by the building code.

There are two interesting reports on damage caused by Hurricane Eloise, which struck the Florida Panhandle the morning of September 23, 1975. One is by Herbert S. Saffir, a Florida consulting engineer; the other is by Bryon Spangler of the University of Florida. The forward movement of the hurricane was unusually fast, causing its duration in a specific area to be lessened, thus minimizing damage from both wind and tidal surge. The still-water height of Panama City was 16 feet above mean sea level, plus about a 3-foot topping wave and wind gusts of 154 mph for a period of one-half hour were measured.

At least one-third of the older structures in the Panama City area collapsed. These were beachfront motels, restaurants, apartments and condominium complexes, and some permanent residences. The structures built on piling survived with minimal damage. In one case, part of a motel was on spread footings and part on piles. Just the part on spread footings was severely damaged.

The anchorage systems, connection between concrete piles or concrete piers and the grade beams under several high-rise buildings were inadequate to resist uplift loads, illustrating that code enforcement or proper inspection by the design engineer is essential.

Many of the residences and some of the buildings were built on spread footings which failed because the sand they were resting on washed away with scour. Failure of the footings resulted in failure of the superstructure.

Some of the high-rise buildings suffered glass damage in both windows and sliding glass doors.

Apparently few, if any, of the residences and buildings were built to conform to South Florida Building Code requirements. (The code was not legally applicable.) If the requirements had been met, much of the damage could have been prevented at a minimum of cost.

Some surprising things were noticed. In almost every case where there was a swimming pool, considerable erosion occurred. Loss of sand beneath the walkways prior to the storm created a channel for the water to flow through and wash out more sand during the storm, which in turn increased both the velocity and quantity of the flow of water in the channel. This ate away the sand supporting adjacent structures, accelerating their failures.

Slabs on grade performed poorly. Often wave action washed out the sand underneath the slab. When this occurred there was no longer support for the structure above it, and failure resulted.

The storm revealed some shoddy construction. Some builders had placed wire mesh for a slab directly on the sand. Then the concrete was poured on top of it, leaving the mesh below and in the sand, where it served no structural purpose. To be effective, it should have been set on blocks or chairs, or pulled up into the slab during the pouring of the concrete.

In some cases cantilevered slabs, for overhangs, were reinforced for the usual downward gravity loads. Unfortunately when waves dashed against the buildings they splashed upward, imposing an upward force against the slab for which it was not reinforced, causing it to crack and fail.

Modular-unit construction: prefabricating the urban shore

The method of building a house, duplex, or larger condominium structure by fabricating modular units in a shop and assembling them at the site is gaining in popularity for development on shoreline property. The larger of these structures are commonly two to three stories in height, and may contain a large number of living units.

Modular construction makes good economic sense, and there is nothing inherently wrong in this approach to coastal construction. These methods have been used in the manufacturing of mobile homes for years although final assembly on mobile homes is done in the shop rather than in the field. Doing as much of the work as possible in a shop can save considerable labor and cost. The workers are not affected by outside weather conditions. They can be paid by piecework, enhancing their productivity. Shop work lends itself to labor-saving equipment such as pneumatic nailing guns and overhead cranes.

If the manufacturer desires it, shop fabrication can permit higher quality. Inspection and control of the whole process are much easier. For instance, there is less hesitation about rejecting a poor piece of lumber when you have a nearby supply of it than if you are building a single dwelling and have just so much lumber on the site.

On the other hand, because so much of the work is done out of the sight of the buyer, there is the opportunity for the manufacturer to take shortcuts if he is so inclined. It is possible that some modular dwelling units have their wiring, plumbing, ventilation, and heating and air conditioning installed at the factory by unqualified personnel, and it is possible the resulting inferior work is either not inspected or inspected by an unconscientious or inept individual. Therefore, it is important to consider the following: Were wiring, plumbing, heating and air conditioning, and ventilation installed at the factory or at the building site? Were the installers licensed and certified? Was the work inspected at both the factory and on the construction?

Most importantly, is the modular dwelling unit built to provide safety in the event of fire? For example, just a few of the many safety features that should be included are two or more exits, stairs remote from each other, masonry fire walls between units, noncombustible wall sheeting, and compartmentalized units so that if fire does occur it will be confined to that one unit.

In general it is very desirable to check the reputation and integrity of the manufacturer just as you would when hiring a contractor to build your individual house on site. The acquisition

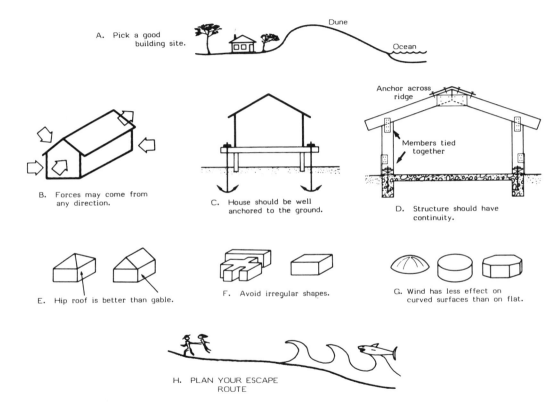

Fig. 6.14. Some rules in selecting or designing a house.

of a modular unit should be approached with the same caution as for other structures.

If you are contemplating purchasing one of these modularized dwelling units, you may be well advised to take the following steps:

1. Check the reputation and integrity of the developer and manufacturer.
2. Check to see if the developer has a state contractor's license.
3. Check the state law on who is required to approve and certify the building.
4. Check what building codes are enforced.
5. Check to see if the state fire marshal's office has indicated that the dwelling units comply with all applicable codes. Also check to see if this office makes periodic inspections.
6. Check to see that smoke alarms have been installed; if windows are the type that can be opened; if the bathroom has an exhaust fan; and if the kitchen has a vent through the roof.

As with all other types of structures, also consider site safety and escape route(s) for the location of modular units.

An unending game: only the players change

Hurricane or calm, receding shore or accreting land, storm-surge flood or sunny sky, migrating dune or maritime forest, win or lose, the gamble of coastal development will continue. If you choose your site with natural safety in view, follow structural engineering design in construction, and take a generally prudent approach to living at the shore (fig. 6.14), then you become the gambler who knows when to hold them, when to fold them, and when to walk away.

Our goal is to provide guidance to today's and tomorrow's players. This book is not the last or by any means the complete guide to coastal living, but it should provide a beginning. In the appendices that follow are additional resources that we hope every reader will pursue.

Appendix A. Hurricane checklist

Keep this checklist handy for protection of family and property.

When a hurricane threatens:

__ Listen for official weather reports.
__ Read your newspaper and listen to radio and television for official announcements.
__ Note the address of the nearest emergency shelter.
__ Know the official evacuation route in advance.
__ Pregnant women, the ill, and infirm should call a physician for advice.
__ Be prepared to turn off gas, water, and electricity where it enters your home.
__ Fill tubs and containers with water (one quart per person per day).
__ Make sure your car's gas tank is full.
__ Secure your boat. Use long lines to allow for rising water.
__ Secure movable objects on your property:
    __ doors
    __ outdoor furniture
    __ shutters
    __ garden tools
    __ hoses
    __ gates
    __ garbage cans
    __ other
__ Board up or tape windows and glassed areas. Draw drapes and window blinds across windows and glass doors. Remove furniture in their vicinity.
__ Stock adequate supplies:
    __ transistor radio
    __ fresh batteries
    __ canned heat
    __ hammer
    __ boards
    __ pliers
    __ hunting knife
    __ tape
    __ first-aid kit
    __ prescribed medicines
    __ water purification tablets
    __ insect repellent
    __ gum, candy
    __ life jackets
    __ charcoal bucket and charcoal
    __ buckets of sand
    __ flashlights
    __ candles
    __ matches
    __ nails
    __ screwdriver
    __ ax*
    __ rope*
    __ plastic drop cloths, waterproof bags, ties
    __ containers for water
    __ disinfectant
    __ canned food, juices, soft drinks (see below)
    __ hard-top head gear
    __ fire extinguisher
    __ can opener and utensils
__ Check mobile-home tiedowns.

*Take an ax (to cut an emergency escape opening) if you go to the upper floors or attic of your home. Take rope for escape to ground when water subsides.

Storm food stock for family of four

— 2 cans, 13-oz. evaporated milk
— 4 cans, 7-oz. fruit juice
— 2 cans tuna, sardines, Spam, chicken
— 3 cans, 10-oz. vegetable soup
— 1 small can of cocoa or Ovaltine
— 1 box, 15-oz. raisins or prunes
— Salt
— Pet food?
— 1 can, 14-oz. cream of wheat or oatmeal
— 1 jar, 8-oz. peanut butter or cheese spread
— 2 cans, 16-oz. pork and beans
— 1 jar, 2-oz. instant coffee or tea bags
— 2 packages of crackers
— 2 pounds of sugar
— 2 quarts of water per person

Special precautions for apartments/condos

— Make one person the building captain to supervise storm preparation.
— Know your exits.
— Count stairs on exits; you'll be evacuating in darkness.
— Locate safest areas for occupants to congregate.
— Close, lock, and tape windows.
— Remove loose items from terraces (and from your absent neighbors' terraces).
— Remove or tie down loose objects from balconies or porches.
— Assume other trapped people may wish to use the building for shelter.

Special precautions for mobile homes

— Pack breakables in padded cartons and place on floor.
— Remove bulbs, lamps, mirrors—put them in the bathtub.
— Tape windows.
— Turn off water, propane gas, electricity.
— Disconnect sewer and water lines.
— Remove awnings.
— **Leave**.

Special precautions for businesses

— Take photos of building and merchandise.
— Assemble insurance policies.
— Move merchandise away from plate glass.
— Move merchandise as high as possible.
— Cover merchandise with tarps or plastic.
— Remove outside display racks and loose signs.
— Take out lower file drawer, wrap in trash bag, and store high.
— Sandbag spaces that may leak.
— Take special precautions with reactive or toxic chemicals.

If you remain at home:

— Never remain in a mobile home; seek official shelter.
— Stay indoors. Remember, the first calm may be the hurricane's eye. Remain indoors until an official all-clear is given.

— Stay on the "downwind" side of the house. Change your position as the wind changes.
— If your house has an "inside" room, it may be the most secure part of the structure.
— Keep continuous communications watch for *official* information on radio and television.
— Keep calm. Your ability to meet emergencies will help others.

If evacuation is advised:

— Leave as soon as you can. Follow official instructions only.
— Follow official evacuation routes unless those in authority direct you to do otherwise.
— Take these supplies:
  — change of warm, protective clothes
  — first-aid kit
  — baby formula
  — identification tags: include name, address, and next of kin (wear them)
  — flashlight
  — food, water, gum, candy
  — rope, hunting knife
  — waterproof bags and ties
  — can opener and utensils
  — disposable diapers
  — special medicine
  — blankets and pillows in waterproof casings
  — radio
  — fresh batteries
  — bottled water
  — can opener and utensils
  — purse, wallet, valuables
  — life jackets
  — games and amusements for children
— Disconnect all electric appliances except refrigerator and freezer. Their controls should be turned to the coldest setting and the doors kept closed.
— Leave food and water for pets. Seeing-eye dogs are the only animals allowed in the shelters.
— Shut off water at the main valve (where it enters your home).
— Lock windows and doors.
— Keep important papers with you:
  — driver's license and other identification
  — insurance policies
  — property inventory
  — Medic Alert or other device to convey special medical information

During the hurricane:

— Stay indoors and away from windows and glassed areas.
— If you are advised to evacuate, **do so at once**.
— Listen for continuing weather bulletins and official reports.
— Use your telephone only in an emergency.

— Follow official instructions only. Ignore rumors.
— Keep **open** a window or door on the side of the house opposite the storm winds.
— Beware the "**eye of the hurricane.**" A lull in the winds is not an indication that the storm has passed. Remain indoors unless emergency repairs are necessary. Exercise caution. Winds may resume suddenly, in the opposite direction and with greater force than before. Remember, if wind direction does change, the open window or door must be changed accordingly.
— Be alert for rising water.
— If electric service is interrupted, note the time.
  — Turn off major appliances, especially air conditioners.
  — Do not disconnect refrigerators or freezers. Their controls should be turned to the coldest setting and doors closed to preserve food as long as possible.
  — Keep away from fallen wires. Report location of such wires to the utility company.
— If you detect **gas**:
  — Do not light matches or electrical equipment.
  — Extinguish all flames.
  — Shut off gas supply at the meter.
  — Report gas service interruptions to the gas company.

Note: Gas should be turned back on only by a gas serviceman or licensed plumber.

— **Water**:
  — The only **safe** water is the water you stored before it had a chance to come in contact with flood waters.
  — Should you require an additional supply, be sure to boil water for 30 minutes before use.
  — If you are unable to boil water, treat water you will need with water purification tablets.

Note: An official announcement will proclaim tap water "safe." Treat all water except stored water until you hear the announcement.

After the hurricane has passed:

— Listen for official word of danger having passed.
— Watch out for loose or hanging power lines as well as gas leaks. People have survived storms only to be accidentally electrocuted or burned. Fire protection may be nil because of broken power lines.
— Walk or drive carefully through the storm-damaged area. Streets will be dangerous because of debris, undermining by washout, and weakened bridges. Watch out for poisonous snakes and insects driven out by flood waters.
— Eat nothing and drink nothing that has been touched by flood waters.
— Place spoiled food in plastic bags and tie securely.
— Dispose of all mattresses, pillows, and cushions that have been in flood waters.
— Contact relatives as soon as possible.

Note: If you are stranded, signal for help by waving a flashlight at night or white cloth during the day.

Appendix B. A guide to federal, state, and local agencies involved in coastal development

Numerous agencies at all levels of government are engaged in planning, regulating, or studying coastal development in Texas. These agencies issue permits for various phases of construction and provide information on development to the homeowner, developer, or planner. Following is an alphabetical list of topics related to coastal development; under each topic are the names of agencies to consult for information on that topic.

Aerial photography, orthophoto maps, and remote-sensing imagery

Persons interested in aerial photography, remote-sensing techniques, or agencies that supply aerial photographs or images should contact:

Texas Natural Resources Information System
P.O. Box 13087
Austin, TX 78711
Phone: (512) 475-3321

Remote Sensing Center
Texas A&M University
College Station, TX 77843

Beach erosion

Information on beach erosion, floods, and storm washover is available from:

Galveston District
U.S. Army Corps of Engineers
P.O. Box 1229
Galveston, TX 77553

Bureau of Economic Geology
The University of Texas at Austin
Box X, University Station
Austin, TX 78712
Phone: (512) 471-1534

Bridges and causeways

The Highway Department has jurisdiction over the building of bridges or causeways. Information is available from:

Texas Department of Highways and Public Transportation
11th and Brazos
Austin, TX 78701
Phone: (512) 475-8044

Appendix B. Guide to agencies

Civil preparedness. See also *Disaster assistance*

For information call or write:

Texas Department of Public Safety
Disaster Emergency Services
5805 N. Lamar Boulevard
Austin, TX 78752
Phone: (512) 465-2138

Coastal Public Lands Management Act, Texas (CPLMA), and Texas Coastal Program (TCP)

The 1973 CPLMA and the TCP both involve land-use planning in coastal counties. The authority of the CPLMA is limited to state-owned lands whereas the TCP designates areas of environmental concern (AECs). These are areas where special environmental, historical, or scientific factors must be considered when development occurs. Permits from the state or from local government (if it chooses to act as a permit-letting agency) are required for the development in some areas. Learn where areas of environmental concern are in relation to your property and what development activities require permits. For further information, write or call:

Texas General Land Office
Stephen F. Austin Building
Austin, TX 78701
Phone: (512) 475-4681

Texas Energy and Natural Resources Advisory Council
200 E. 18th Street
Austin, TX 78701
Phone: (512) 475-0414

Office of Coastal Zone Management
National Oceanic and Atmospheric Administration
3300 Whitehaven Street, N.W.
Washington, D.C. 20235
Phone: (202) 634-6791
(provides publications to aid planners and managers)

Disaster assistance

For information, call or write:

Texas Department of Public Safety
Disaster Emergency Services
5805 N. Lamar Boulevard
Austin, TX 78752
Phone: (512) 465-2138

Federal Emergency Management Agency
NFIP Region VI: Texas
Federal Center
Denton, TX 76201
Phone: (817) 387-5811

American National Red Cross
Disaster Services
Washington, D.C. 20006
Phone: (202) 857-3722

Dredging, filling, and construction in coastal waterways

Texas law requires that all those who wish to dredge, fill, or otherwise alter marshlands, estuarine bottoms, or tidelands apply for a permit from the General Land Office. For information, write or call:

Texas General Land Office
Stephen F. Austin Building
Austin, TX 78701
Phone: (512) 475-4681

Federal law requires that any person who wishes to dredge, fill, or place any structure in navigable water (almost any body of water) apply for a permit from the U.S. Army Corps of Engineers. Information is available from:

Galveston District
U.S. Army Corps of Engineers
P.O. Box 1229
Galveston, TX 77553

Dune alteration

Texas coastal counties have the option of controlling the destruction, damage, or removal of any sand dune or dune vegetation. Individual counties may have ordinances pertaining to dune alteration. Permits for certain types of alteration may be obtained from the local dune-protection board. For information, call or write the local county courthouse.

Geologic information

For information call or write:

Branch of Distribution
U.S. Geological Survey
1200 South Eads Street
Arlington, VA 22202
(Request Geologic and Water-Supply Reports and Maps, Texas; free index.)

Bureau of Economic Geology
The University of Texas at Austin
Box X, University Station
Austin, TX 78712
Phone: (512) 471-1534

Hazards. See also *Beach erosion* and *Insurance*

Literature describing natural hazards is available from:

Texas Coastal and Marine Council
P.O. Box 13407
Austin, TX 78711
Phone: (512) 475-5849

Bureau of Economic Geology
The University of Texas at Austin
Box X, University Station
Austin, TX 78712
Phone: (512) 471-1534

Office of Coastal Zone Management
National Oceanic and Atmospheric Administration
3300 Whitehaven Street, N.W.
Washington, D.C. 20235

Sea Grant College Program
Texas A&M University
College Station, TX 77843
Phone: (713) 845-3854

Health. See also *Sanitation*

The local Department of Health is in charge of issuing home septic-tank permits. Questions may be directed to the officer of the local agency or to:

Texas Department of Health
1100 W. 49th Street
Austin, TX 78756
Phone: (512) 458-7111

History/archaeology

For information call or write:

Texas Historical Commission
1151 Colorado
Austin, TX 78701
Phone: (512) 475-3092

Housing. See *Subdivisions*

Hurricane awareness. See *appendix C*

Hurricane information

The National Oceanic and Atmospheric Administration (NOAA) is the best agency from which to request information on hurricanes. NOAA Storm-Evacuation Maps are prepared for vulnerable areas and are available for minimal cost. To find out whether a map is available for your area, call or write:

Distribution Division (C-44)
National Ocean Survey
National Oceanic and Atmospheric Administration
Riverdale, MD 20840
Phone: (301) 436-6990

Texas Coastal and Marine Council
P.O. Box 13407
Austin, TX 78711
Phone: (512) 475-5849

Sea Grant College Program
Texas A&M University
College Station, TX 77843
Phone: (713) 845-3854

Insurance

In coastal areas special building requirements must often be met in order to obtain flood or windstorm insurance. To find out the requirements for your area, check with your insurance agent.

Director
Office of National and Technological Hazards Division
Federal Emergency Management Agency
NFIP Region VI: Texas
Federal Center
Denton, TX 76201
Phone: (817) 387-5811

Insurance Information Institute
1011 Congress Avenue, Suite 501
Austin, TX 78701

For V-Zone Coverage contact:

NFIP
Attn.: V-Zone Underwriting Specialist
6430 Rockledge Drive
Bethesda, MD 20817
Phone: (800) 638-6620

Land acquisition

When acquiring property or a condominium—whether in a subdivision or not—consider the following: (1) Owners of property next to dredged canals should make sure that the canals are designed for adequate flushing to keep the canals from becoming stagnant. (2) Description and survey of land in coastal areas are very complicated. Old titles granting fee-simple rights to property below the high-tide line may not be upheld in court; titles should be reviewed by a competent attorney before they are transferred. A boundary described as the high-water mark may be impossible to determine. (3) Ask about the provision of sewage disposal and utilities including water, electricity, gas, and telephone. (4) Be sure any promises of future improvements, access, utilities, additions, common property rights, etc., are in writing. (5) Be sure to visit the property and inspect it carefully before buying it. (See the following sections on Planning and land use and on Subdivisions.)

Maps

A wide variety of maps is useful to planners and managers and may be of interest to individual property owners. Topographic, geologic, and land-use maps are available from:

Distribution Section
U.S. Geological Survey
1200 South Eads Street
Arlington, VA 22202

Appendix B. Guide to agencies

Bureau of Economic Geology
The University of Texas at Austin
Box X, University Station
Austin, TX 78712
Phone: (512) 471-1534

Evacuation maps: See *Hurricane information*
Flood-zone maps: See *Insurance*
Planning maps: Call or write your local county commission.
Nautical charts in several scales contain navigation information on Texas coastal waters. A nautical-chart index map is available from:

National Ocean Survey
Distribution Division (C-44)
National Oceanic and Atmospheric Administration
Riverdale, MD 20840
Phone: (301) 436-6990

County-highway base maps are available from:

Department of Highways and Public Transportation
11th and Brazos
Austin, TX 78701

Marine and coastal-zone information

Texas has several marine advisory centers. Located conveniently along the coast, the centers are designed to help residents with marine related questions. For further information contact:

Sea Grant College Program
Texas A&M University
College Station, TX 77843
Phone: (713) 845-3854

Texas Coastal and Marine Council
P.O. Box 13407
Austin, TX 78711
Phone: (512) 475-5849

Parks and recreation

Texas Parks and Wildlife Department
4200 Smith School Road
Austin, TX 78744
Phone: (512) 475-4888

Padre Island National Seashore
9405 South Padre Island Drive
Corpus Christi, TX 78418

Planning and land use. See also *Coastal Public Lands Management Act, Texas*

For information call or write:

Texas General Land Office
Stephen F. Austin Building
Austin, TX 78701
Phone: (512) 475-4681

For specific information on your area, check with your local town or county commission. Many local governments have planning boards that answer to the commission and have available copies of existing or proposed land-use plans.

Roads: public rights and beach access

Texas Attorney General's Office
P.O. Box 12548
Austin, TX 78711
Phone: (512) 475-2501

The Department of Highways and Public Transportation is not required to furnish access to all property owners. Before buying property, make sure that access rights and roads will be provided.

Permits to connect driveways from commercial developments to state-maintained roads must be obtained from the district engineer of the Highway Department.

Subdivision roads—privately built roads including those serving seasonal residences—must meet specific standards in order to be eligible for addition to the state highway system. Information on these requirements may be obtained from the district engineer of the Highway Department. Further information is available from local offices.

Sanitation

Before construction permits will be issued, improvement permits for septic tanks must be obtained from the local Department of Health. Improvement permits are based on soil suitability for septic-tank systems, and apply to conventional homes and mobile homes outside of mobile-home parks, in areas that are not serviced by public or community sewage systems.

A permit for the construction of a sewage-disposal system or any other structure in navigable waters must be obtained from the U.S. Army Corps of Engineers. More information is available from:

Galveston District
U.S. Army Corps of Engineers
P.O. Box 1229
Galveston, TX 77553

A permit for any discharge into navigable waters must be obtained from the U.S. Environmental Protection Agency. Recent judicial interpretation of the Federal Water Pollution Control Amendments of 1972 extends above the mean high-water mark federal jurisdiction for protection of wetland. Federal permits may now be required for the development of land that occasionally is flooded by water draining indirectly into a navigable waterway.

Soils

Soil type is important in terms of (1) the type of vegetation it can support, (2) the type of construction technique it can withstand, (3) its drainage characteristics, and (4) its ability to accommodate septic systems. Ultimately, soil reports including maps and rating of soil types will be available for most of the Texas coast. The following agency produces a variety of reports that are useful to property owners:

> U.S. Department of Agriculture
> Soil Conservation Service
> 1106 Clayton Lane
> Austin, TX 78723
> Phone: (512) 397-5592

Subdivisions

Subdivisions containing more than 100 lots and offered in interstate commerce must be registered with the Office of Interstate Land Sales Registration (as specified by the Interstate Land Sales Full Disclosure Act). Prospective buyers must be provided with a property report. This office also produces a booklet entitled *Get the Facts . . . Before Buying Land* for people who wish to invest in land. Information on subdivision property and land investment is available from:

> Office of Interstate Land Sales Registration
> U.S. Department of Housing and Urban Development
> Washington, D.C. 20410

Vegetation

Information on vegetation may be obtained from the local Soil and Water Conservation district office. For information on the use of grass and other plantings for stabilization or aesthetics, consult the publications listed in appendix C under Vegetation.

Water resources

Several agencies are concerned with water quality and availability. The following ones will answer questions on this subject:

> Texas Department of Water Resources
> 1700 North Congress
> Austin, TX 78701
> Phone: (512) 475-7036

> U.S. Geological Survey
> Water Resources Division
> 300 East 8th Street
> Austin, TX 78701
> Phone: (512) 397-5766

Weather

Hurricane information is available from:

> National Weather Service
> Southern Region
> 819 Taylor Street
> Fort Worth, TX 76102

For current weather information, listen to your local radio and television stations.

Appendix C. Useful references

The following publications are listed by subject; subjects are arranged in the approximate order that they appear in the preceding chapters. A brief description of each reference is provided, and sources are included for those readers who would like to obtain more information on a particular subject. Most of the references listed are either low in cost or free or available in major libraries; we encourage the reader to take advantage of these informative publications.

History

1. *The Early History of Galveston*, by J. O. Dyer, 1916. Centenary edition, part 1, Oscar Springer, Galveston, TX. Available at university libraries in Texas.
2. *Historical Sketch of the Explorations in the Gulf of Mexico*, by P. S. Galtsoff, 1954. Published in the *U.S. Fish and Wildlife Service Bulletin* 89, pp. 3-36. Available in large university libraries.
3. *Treasure, People, Ships, and Dreams*, by J. L. Davis, 1977. Summarizes Spanish exploration and occupation of the Texas coast and describes the restoration of artifacts salvaged from one of the shipwrecks. Published by the Texas Antiquities Committee, Box 12276 Capitol Station, Austin, TX 78711.
4. *Annual Report of Chief of Engineers*, by U.S. Army Corps of Engineers, 1850-present. Summarizes activities related to jetty construction, dredging, and harbor improvements. Published in House and Senate documents. Available only at Federal repositories and Corps of Engineers District offices.
5. *Padre Island National Seashore Historic Resources Study*, by J. W. Sheire, 1971. U.S. Department of Interior, National Park Service, 9405 South Padre Island Drive, Corpus Christi, TX 78418.

Hurricanes

6. *Early American Hurricanes, 1492-1870*, by D. M. Ludlum, 1963. Informative and entertaining descriptions of storms affecting the Atlantic and Gulf coasts. Storm accounts in chronological order provide insight into the frequency, intensity, and destructive potential of hurricanes. Published by the American Meteorological Society, Boston, MA. Available in public and university libraries.
7. *Hurricanes as Geological Agents: Case Studies of Hurricanes Carla, 1961, and Cindy, 1963*, by M. O. Hayes, 1967; reprinted 1974. Describes the response of Texas beaches and barriers to two different storms. Available from the Bureau of Economic Geology, Box X, University Station, Austin, TX 78712.

8. *Atlantic Hurricanes*, by G. E. Dunn and B. I. Miller, 1960. Discusses at length hurricanes and associated phenomena such as storm surge, wind, and sea action. Includes a detailed account of Hurricane Hazel, 1954, and suggestions for pre- and posthurricane procedures. An appendix includes a list of hurricanes for the Gulf of Mexico. Published by the Louisiana State University Press, Baton Rouge, LA 70893. Available in public and college libraries.

9. *Hurricane Information and Gulf Tracking Chart*, by the National Oceanic and Atmospheric Administration, 1974. A brochure that describes hurricanes, defines terms, and lists hurricane safety rules. Outlines methods of tracking hurricanes and provides a tracking map. Available from the Superintendent of Documents, U.S. Government Printing Office, Washington, D.C. 20402.

10. *Hurricanes on the Texas Coast*, by Walter Henry, D. N. Driscoll, and J. P. McCormack, 1975. Coastal resident's guide to understanding, preparing for, and recovering from the effects of hurricanes. Available at no charge from Sea Grant College Program, Texas A&M University, College Station, TX 77843 (Publication TAMU-SG-75-504).

11. *Hurricane Surge Frequency Estimated for the Gulf Coast of Texas*, by B. R. Bodine, 1969. Technical Memorandum 26 of the Coastal Engineering Research Center, U.S. Army Corps of Engineers, 5201 Little Falls Road, N.W., Washington, D.C. 20016.

12. *A Weekend in September*, by John E. Weems, 1980. An historical account of the Galveston Hurricane of 1900 originally researched and written in the 1950s when survivors of the great tragedy were still living and available for interviews. The eyewitness accounts and the study of contemporary documents are a firm basis for a readable account of this greatest of all Texas storms. Published by Texas A&M University Press, College Station, TX 77843.

The following three references (13, 14, and 15) are individual reports on the three most recent storms that struck the Texas coast. They contain photographs, maps of flooded areas, damage estimates, and descriptions of the storms. Available from the Galveston District, U.S. Army Corps of Engineers, P.O. Box 1229, Galveston, TX 77553.

13. *Report on Hurricane Carla, 9-12 September 1961*, by the U.S. Army Corps of Engineers, Galveston District, 1962.

14. *Report on Hurricane Beulah, 8-12 September 1967*, by the U.S. Army Corps of Engineers, Galveston District, 1968.

15. *Report on Hurricane Celia, 30 July-5 August 1970*, by the U.S. Army Corps of Engineers, Galveston District, 1970.

16. *Hurricanes Affecting the Coast of Texas from Galveston to the Rio Grande*, by W. A. Price, 1956. Printed as U.S. Army Corps of Engineers Erosion Board Technical Memorandum 78. Available at the Galveston District, Corps of Engineers.

17. *Texas Coastal Hurricane Study, Volume 1*, by the U.S. Army Corps of Engineers, 1979. Available from the Galveston District, U.S. Army Corps of Engineers, P.O. Box 1229, Galveston, TX 77553.
18. *Atlantic Hurricane Frequencies along the U.S. Coastline*, by R. H. Simpson and M. B. Lawrence, 1971. National Oceanic and Atmospheric Administration (NOAA) Technical Memorandum NWS SR-58. Available in large university libraries.

Hurricane awareness

19. *Hurricane Awareness.* An outstanding set of pamphlets outlining what happens when a hurricane strikes, the difference between a "watch" and a "warning," and how to track a hurricane, and providing a significant hurricane checklist. The sheets also include maps of general flood zones, areas flooded by specific hurricanes, and a tracking chart. Information is provided on insurance and storm recovery. The motto "advance preparation is the key to saving lives and property" says it best, and we encourage our coastal readers to request the pamphlet sheet for their area from: "Hurricane Awareness," Texas Coastal and Marine Council, Box 13407, Austin, TX 78711. Pamphlets are available for the following areas: Bay City area; Beaumont area; Brownsville area; Corpus Christi area; Galveston area; Houston area; Kingsville area; Port Lavaca area; Brazoria, Galveston, and South Harris Counties.
20. *Hurricane Watch... Hurricane Warning: Why Don't People Listen?* by Carlton Ruch and Larry Christensen, 1980. Available at no charge from Sea Grant College Program, Texas A&M University, College Station, TX 77843 (Publication TAMU-SG-80-508).

Storm survival

21. *Surviving the Storm*, 1981. An excellent twelve-page summary of storm precautions prepared for the Virgin Islands but useful anywhere. Available from "Surviving the Storm," P.O. Box 1208, St. Thomas, U.S. Virgin Islands 00801.

Geology and land use

Each of the following seven references (22-28) is an atlas for a particular section of the Texas coast, and includes nine multicolored geologic and environmental maps accompanied by text explaining their use and interpretation. Taken together, the seven atlases comprise a folio publication entitled *Environmental Geologic Atlas of the Texas Coastal Zone*. Atlases for each of the areas can be purchased separately. Available from the Bureau of Economic Geology, Box X, University Station, Austin, TX 78712.

22. *Bay City-Freeport Area*, by J. H. McGowen, L. F. Brown, Jr., T. J. Evans, W. L. Fisher, and C. G. Groat, 1976.
23. *Beaumont-Port Arthur Area*, By W. L. Fisher, L. F. Brown, Jr., J. H. McGowen, and C. G. Groat, 1973.
24. *Brownsville-Harlingen Area*, by L. F. Brown, Jr., J. L. Brewton, T. J. Evans, J. H. McGowen, W. A. White, C. G. Groat, and W. L. Fisher, 1980.

25. *Corpus Christi Area*, by L. F. Brown, Jr., J. L. Brewton, J. H. McGowen, T. J. Evans, W. L. Fisher, and C. G. Groat, 1976.
26. *Galveston-Houston Area*, by W. L. Fisher, J. H. McGowen, L. F. Brown, Jr., and C. G. Groat, 1972.
27. *Kingsville Area*, by L. F. Brown, Jr., J. H. McGowen, T. J. Evans, C. G. Groat, and W. L. Fisher, 1977.
28. *Port Lavaca Area*, by J. H. McGowen, C. V. Proctor, Jr., L. F. Brown, Jr., T. J. Evans, W. L. Fisher, and C. G. Groat, 1976.
29. *Resource Capability Units—Their Utility in Land- and Water-Use Management with Examples from the Texas Coastal Zone*, by L. F. Brown, Jr., W. L. Fisher, A. W. Erxleben, and J. H. McGowen, 1971. This twenty-two page report provides definitions and examples of the physical properties, biologic assemblages, and active processes that are of first-order importance in any management plan. Available from the Bureau of Economic Geology, Box X, University Station, Austin, TX 78712.

Barrier islands

30. *Barrier Islands and Beaches*, 1976. Proceedings of the May 1976 barrier-islands workshop. A collection of technical papers prepared by scientists studying islands. Provides an up-to-date, readable overview of barrier islands. Comprehensive coverage—from aesthetics to flood insurance—by experts. Topics include island ecosystems, ecology, geology, politics, and planning. Good bibliographic source for those studying barrier islands. Available from the Publications Department, Conservation Foundation, 1717 Massachusetts Avenue, N.W., Washington, D.C. 20036.
31. *Land and Water Resources, Historical Changes, and Dune Criticality, Mustang and North Padre Island, Texas*, by W. A. White, R. A. Morton, R. S. Kerr, W. D. Kuenzi, and W. B. Brogden, 1978. Available from the Bureau of Economic Geology, Box X, University Station, Austin, TX 78712.
32. *Barrier Island Formation*, by J. H. Hoyt, 1967. A technical paper in which Hoyt develops his theory of barrier-island formation. Published in the *Bulletin of the Geological Society of America*, v. 78, pp. 1125-1136. Available only in the larger university libraries.
33. *Coastal Geomorphology*, edited by D. R. Coates, 1973. Another collection of technical papers including R. Dolan's "Barrier Islands: Natural and Controlled," and P. J. Godfrey and M. M. Godfrey's "Comparison of Ecological and Geomorphic Interaction between Altered and Unaltered Barrier Island Systems in North Carolina." Interesting reading for anyone willing to overlook the jargon of coastal scientists. Published by the State University of New York, Binghamton, NY 13901. Available only in the larger university libraries.
34. *Effects of Hurricane Celia—A Focus on Environmental Geologic Problems of the Texas Coastal Zone*, by J. H. McGowen, C. G. Groat, L. F. Brown, Jr., W. L. Fisher, and

A. J. Scott, 1970. Available from the Bureau of Economic Geology, Box X, University Station, Austin, TX 78712.

Barrier-island environments

35. *Modern Depositional Environments of the Texas Coast*, by R. A. Morton and J. H. McGowen, 1980. A guidebook describing the physical characteristics and processes of coastal environments. Available from the Bureau of Economic Geology, Box X, University Station, Austin, TX 78712.
36. *Padre Island National Seashore—A Guide to the Geology, Natural Environments, and History of a Texas Barrier Island*, by B. R. Weise and W. A. White, 1980. Available from the Bureau of Economic Geology, Box X, University Station, Austin, TX 78712.

Beaches

37. *Waves and Beaches*, by Willard Bascom, 1964. A discussion of beaches and coastal processes. Published by Anchor Books, Doubleday and Co., Garden City, NY 11530. Available in paperback from local bookstores.
38. *Beaches and Coasts*, 2nd edition, by C. A. M. King, 1972. Classic treatment of beach and coastal processes. Published by St. Martin's Press, Inc., 175 Fifth Avenue, New York, NY 10010.
39. *Beach Processes and Sedimentation*, by Paul Komar, 1976. The most up-to-date technical explanations of beaches and beach processes. Recommended only to serious students of the beach. Published by Prentice-Hall, Englewood Cliffs, NJ 07632.
40. *Land Against the Sea*, by the U.S. Army Corps of Engineers, 1964. Readable introduction to coastal geology and shoreline processes. The belief of authors in the value of certain engineering methods, however, is either outdated or unsubstantiated. Available as *Miscellaneous Paper No. 4-64* from the U.S. Army Corps of Engineers, Coastal Engineering Research Center, Kingman Building, Ft. Belvoir, VA 22060.

Recreation

41. *Recreation in the Coastal Zone*, 1975. A collection of papers presented at a symposium sponsored by the U.S. Department of the Interior, Bureau of Outdoor Recreation, Southeast Region. Outlines different views of recreation in the coastal zone and the approaches taken by some states to recreation-related problems. The symposium was co-sponsored by the Office of Coastal Zone Management. Available from that office, National Oceanic and Atmospheric Administration, 3300 Whitehaven Street, N.W., Washington, D.C. 20235.
42. *Beachcomber's Guide to Gulf Coast Marine Life*, by Nick Fotheringham, 1980, Gulf Publishing Company, Houston, TX.

Appendix C. Useful references

43. *A Birder's Guide to the Texas Coast*, by J. A. Lane and J. L. Tveten, 1980. Available from P.O. Box 21604, Denver, CO 80221.
44. *Saltwater Fishes of Texas*, 1970. Published as *Bulletin No. 52*, Texas Parks and Wildlife, 4200 Smith School Road, Austin, TX 78744.
45. *Seashells of the Texas Coast*, by Jean Andrews, 1971. Illustrates and describes the common seashells found on the beach. Published by The University of Texas Press, Austin, TX 78712.
46. *Design of Inlets for Texas Coastal Fisheries*, by H. P. Carothers and H. C. Innis, 1962. A technical paper that describes the plan and rationale for cutting water exchange or fish passes across the Texas barrier islands. *American Society of Civil Engineers Transactions*, v. 127, pp. 231-259. Available only in the larger university libraries.
47. *An Economic Inventory of Recreation and Tourism within the Texas Coastal Zone*, by Billie Ingram, 1973. Available from Sea Grant College Program, Texas A&M University, College Station, TX 77843 (Publication TAMU-SG-73-209).
48. *A Recreational Guide to the Central Texas Coast*, by Edwin Doran, Jr., and B. P. Brown, 1975. Describes a wide variety of recreational facilities. Available at no charge from Sea Grant College Program, Texas A&M University, College Station, TX 77843 (Publication TAMU-SG-75-606).

Shoreline engineering

49. *Shore Protection Guidelines*, by the U.S. Army Corps of Engineers, 1971. Summary of the effects of waves, tides, and winds on beaches and engineering structures used for beach stabilization. Available free from the Department of the Army, Corps of Engineers, Washington, D.C. 20318.
50. *Shore Protection Manual*, by the U.S. Army Corps of Engineers, 1973. The "bible" of shoreline engineering. Published in three volumes. Request publication 08-0-22-0007 from the Superintendent of Documents, U.S. Government Printing Office, Washington, D.C. 20402.
51. *Help Yourself*, by the U.S. Army Corps of Engineers. Brochure addressing the erosion problems in the Great Lakes region. May be of interest to barrier-island residents as it outlines shoreline processes and illustrates a variety of shoreline-engineering devices used to combat erosion. Free from the U.S. Army Corps of Engineers, North Central Division, 219 South Dearborn Street, Chicago, IL 60604.
52. *Shore Protection*, by J. B. Herbich and R. E. Schiller, Jr., 1976. Advisory bulletin. Available at no charge from Sea Grant College Program, Texas A&M University, College Station, TX 77843 (Publication TAMU-SG-76-504).
53. *Coastal Hydraulics*, by A. M. Muir Wood, 1969. A shoreline-engineering textbook suitable for engineering students. Published by Gordon and Breach Science Publishers, Inc., 150 Fifth Avenue, New York, NY 10011.

54. *Publications List, Coastal Engineering Research Center (CERC) and Beach Erosion Board (BEB)*, by the U.S. Army Corps of Engineers, 1976. A list of published research by the U.S. Army Corps of Engineers. Free from the U.S. Army Corps of Engineers, Coastal Engineering Research Center, Kingman Building, Ft. Belvoir, VA 22060.

55. *The Gulf Shoreline of Texas: Processes, Characteristics, and Factors in Use*, by J. H. McGowen, L. E. Garner, and B. H. Wilkinson, 1977. Available from the Bureau of Economic Geology, Box X, University Station, Austin, TX 78712.

56. *Galveston's Bulwark Against the Sea, History of the Galveston Seawall*, by A. B. Davis, Jr., 1974. The story of the Galveston seawall illustrates how the structure has been enlarged with time. If still in print, it is available from the Public Affairs Officer, Galveston District, U.S. Army Corps of Engineers, P.O. Box 1229, Galveston, TX 77553.

Hazards

57. *Natural Hazards of the Texas Coastal Zone*, by L. F. Brown, Jr., R. A. Morton, J. H. McGowen, C. W. Kreitler, and W. L. Fisher, 1974. A series of maps with accompanying text that summarizes the areas influenced by hurricane flooding, shoreline erosion, land surface subsidence, and active faulting. Available from the Bureau of Economic Geology, Box X, University Station, Austin, TX 78712.

58. *Pictorial Atlas of Texas Coastal Hazards*, by the Texas Coastal and Marine Council, 1977. Striking photographic summary of coastal hazards which heightens one's awareness. Available from the Texas Coastal and Marine Council, P.O. Box 13407, Austin, TX 78711.

The following eight references (59-66) are a series of eight circulars that present detailed measurements of shoreline changes between the late 1800s and the mid-1970s. Available from the Bureau of Economic Geology, Box X, University Station, Austin, TX 78712.

59. *Shoreline Changes Between Sabine Pass and Bolivar Roads, An Analysis of Historical Changes of the Texas Gulf Shoreline*, by R. A. Morton, 1975 (Geological Circular 75-6).

60. *Shoreline Changes on Galveston Island (Bolivar Roads to San Luis Pass), An Analysis of Historical Shoreline Changes of the Texas Gulf Shoreline*, by R. A. Morton, 1974 (Geological Circular 74-2).

61. *Shoreline Changes in the Vicinity of the Brazos River Delta (San Luis Pass to Brown Cedar Cut), An Analysis of Historical Changes of the Texas Gulf Shoreline*, by R. A. Morton and M. J. Pieper, 1975 (Geological Circular 75-4).

62. *Shoreline Changes on Matagorda Peninsula (Brown Cedar Cut to Pass Cavallo), An Analysis of Historical Changes of the Texas Gulf Shoreline*, by R. A. Morton, M. J. Pieper, and J. H. McGowen, 1976 (Geological Circular 76-6).

63. *Shoreline Changes on Matagorda Island and San Jose Island (Pass Cavallo to Aransas Pass), An Analysis of Historical*

Changes of the Texas Gulf Shoreline, by R. A. Morton and M. J. Pieper, 1976 (Geological Circular 76-4).

64. *Shoreline Changes on Mustang Island and North Padre Island (Aransas Pass to Yarborough Pass)—An Analysis of Historical Changes of the Texas Gulf Shoreline,* by R. A. Morton and M. J. Pieper, 1977 (Geological Circular 77-1).

65. *Shoreline Changes on Central Padre Island (Yarborough Pass to Mansfield Channel)—An Analysis of Historical Changes of the Texas Gulf Shoreline,* by R. A. Morton and M. J. Pieper, 1977 (Geological Circular 77-2).

66. *Shoreline Changes on Brazos Island and South Padre Island (Mansfield Channel to Mouth of the Rio Grande), An Analysis of Historical Changes of the Texas Gulf Shoreline,* by R. A. Morton and M. J. Pieper, 1975 (Geological Circular 75-2).

67. *Historical Shoreline Changes and Their Causes, Texas Gulf Coast,* by R. A. Morton, 1977. Geological Circular 77-6. Available from the Bureau of Economic Geology, Box X, University Station, Austin, TX 78712.

68. *Historical Changes and Related Coastal Processes, Gulf and Mainland Shorelines, Matagorda Bay Area, Texas,* by J. H. McGowen and J. L. Brewton, 1975. Available from the Bureau of Economic Geology, Box X, University Station, Austin, TX 78712.

69. *Investigation of Shoreline Changes at Sargent Beach, Texas,* by W. N. Seelig and R. M. Sorensen, 1973. Available from Sea Grant College Program, Texas A&M University, College Station, TX 77843 (Publication TAMU-SG-73-212).

70. *Quantitative Analysis of Shoreline Change, Sargent, Texas,* by J. E. Sealey, Jr., and W. M. Ahr, 1975. Available from Sea Grant College Program, Texas A&M University, College Station, TX 77843 (Publication TAMU-SG-75-209).

71. *Lineations and Faults in the Texas Coastal Zone,* by C. W. Kreitler, 1976. Documents recent movement and structural damage associated with fault activation. Report of Investigations 85. Available from the Bureau of Economic Geology, Box X, University Station, Austin, TX 78712.

72. *Natural Hazard Management in Coastal Areas,* by G. F. White and others, 1976. The most recent summary of coastal hazards along the entire U.S. coast. Discusses adjustments to such hazards and hazard-related federal policy and programs. Summarizes hazard-management and coastal-land-planning programs in each state. Appendices include a directory of agencies, an annotated bibliography, and information on hurricanes. An invaluable reference, recommended to developers, planners, and managers. Available from the Office of Coastal Zone Management, National Oceanic and Atmospheric Administration, 3300 Whitehaven Street, N.W., Washington, D.C. 20235.

73. *Guidelines for Identifying Coastal High Hazard Zones,* by the U.S. Army Corps of Engineers, 1975. Report outlining such zones with emphasis on "coastal special flood-hazard

areas" (coastal flood plains subject to inundation by hurricane surge with one percent chance of occurring in any given year). Provides technical guidelines for conducting uniform flood-insurance studies and outlines methods of obtaining 100-year-storm-surge elevations. Recommended to island planners. Available from the Galveston District, U.S. Army Corps of Engineers, P.O. Box 1229, Galveston, TX 77553.

74. *Galveston Island—A Changing Environment*, by A. R. Benton, Jr., and others, 1979. Remote-sensing study which includes information on beach erosion, the impact of construction, and other factors related to island development. Available from Sea Grant College Program, Texas A&M University, College Station, TX 77843 (Publication TAMU-SG-80-201).

75. *Hazard Awareness Guidebook: Planning for What Comes Naturally*, by Sally Davenport and Penny Waterstone, 1979. Available from the Texas Coastal and Marine Council, P.O. Box 13407, Austin, TX 78711.

Vegetation

The following two reports describe methods and results of using various grasses to stabilize and build dunes. Available from the Gulf Universities Research Consortium, Bellaire, TX 77401.

76. *The Use of Grasses for Dune Stabilization Along the Gulf Coast With Initial Emphasis on the Texas Coast*, by L. C. Otteni and R. L. Baker, 1972. *Gulf Universities Research Consortium Report No. 120.*

77. *Stabilization and Reconstruction of Texas Coastal Foredunes With Vegetation*, by B. E. Dahl, A. Lohse, and S. G. Appan, 1974. *Gulf Universities Research Consortium Report No. 139.*

78. *Flora of the Texas Coastal Bend*, by F. B. Jones, 1977. Published as Welder Wildlife Foundation Contribution B-6. Available in university and public libraries near the coast.

Site analysis

79. *Handbook: Building in the Coastal Environment*, by R. T. Segrest and Associates, 1975. A well-illustrated, clearly and simply written book on Georgia coastal-zone planning, construction, and selling problems. Topics include vegetation, soil, drainage, setback requirements, access, climate, and building orientation. Includes a list of addresses for agencies and other sources of information. Much of the information applies to Texas. Available from the Graphics Department, Coastal Area Planning and Development Commission, P.O. Box 1316, Brunswick, GA 31520.

Water problems

80. *Your Home Septic System, Success or Failure.* Brochure providing answers to commonly asked questions on home septic systems. Lists agencies that supply information on

septic-tank installation and operation. Available from the UNC Sea Grant, 1235 Burlington Laboratories, North Carolina State University, Raleigh, NC 27607.

81. *Evaluation of Sanitary Landfill Sites, Texas Coastal Zone—Geologic and Engineering Criteria*, by L. F. Brown, Jr., W. L. Fisher, and J. F. Malina, Jr., 1972. Geological Circular 72-3. Available from the Bureau of Economic Geology, Box X, University Station, Austin, TX 78712.

82. *Report of Investigation of the Environmental Effects of Private Waterfront Lands*, by W. Barada and W. M. Partington, 1972. An enlightening reference which treats the effects of finger canals on water quality. Available from the Environmental Information Center, Florida Conservation Foundation, Inc., 935 Orange Avenue, Winter Park, FL 32789.

Conservation and planning

83. *The Water's Edge: Critical Problems of the Coastal Zone*, edited by B. H. Ketchum, 1972. The best available scientific summary on coastal-zone problems. Published by the M.I.T. Press, Cambridge, MA 02139.

84. *Texas Coastal Management Program*, prepared by the General Land Office of Texas, 1976. A three-volume summary of the proposed activity assessment routine and other pertinent information relating to coastal-zone management activities. Available from the Texas General Land Office, Austin, TX 78701.

85. *Design with Nature*, by Ian McHarg, 1969. A now-classic text on the environment. Stresses that when man interacts with nature, he must recognize its processes and governing laws, and realize that it both presents opportunities for and requires limitations on human use. Published by Doubleday and Company, Inc., Garden City, NY 11530.

86. *Coastal Ecosystems, Ecological Considerations for Management of the Coastal Zone*, by John Clark, 1974. A clearly written, well-illustrated book on the applications of the principles of ecology to the major coastal-zone environments. Available from the Publications Department, The Conservation Foundation, 1717 Massachusetts Avenue, N.W., Washington, D.C. 20036.

87. *Who's Minding the Shore*, by the Natural Resources Defense Council, Inc., 1976. A guide to public participation in the coastal-zone management processes. Defines coastal ecosystems and outlines the Coastal Zone Management Act, coastal-development issues, and means of citizen participation in the coastal-zone management process. Lists sources of additional information. Available from the Office of Coastal Zone Management, National Oceanic and Atmospheric Administration, 3300 Whitehaven Street, N.W., Washington, D.C. 20235.

88. *The Fiscal Impact of Residential and Commercial Development. A Case Study*, by T. Muller and G. Dawson, 1972. A classic study which demonstrates that development may ulti-

mately increase, rather than decrease, community taxes. Available from the Publications Office, Urban Institute, 2100 M Street, N.W., Washington, D.C. 20037. Refer to URI-22000 when ordering.

Regulation

89. *State Hearing Draft, Texas Coastal Program*, September 1980. This is the document which would have guided coastal land use if Texas had remained in the Federal Coastal Zone Management Program. For a thorough treatment of coastal issues obtain a copy of the draft from the Natural Resources Division, Texas Energy and Natural Resources Advisory Council, 200 E. 18th Street, Austin, TX 78701.

90. *The Law of the Coast in a Clamshell, Part V: The Texas Approach*, by Peter H. F. Graber, 1981. An excellent review of Texas coastal legislation and how it has merged principles from both civil and common law. The paper treats title to lands, determination of tidal boundaries, public trust doctrine, public access rights, private littoral rights, and leasing and regulation of coastal-zone lands and waters. Published in *Shore and Beach*, v. 49, no. 4, pp. 24-31. Available from most university libraries.

91. *Comparative Aspects of Coastal Zone Management Background Information on the Law of Texas and Other States in View of the Coastal Zone Management Act of 1972*, by Carol Dinkins, 1973. Available from Sea Grant College Program, Texas A&M University, College Station, TX 77843 (Publication TAMU-SG-73-607).

Building a new home

Both current and prospective owners and builders of homes in hurricane-prone areas should supplement the information and advice provided in this book with that offered in references dealing specifically with safe construction. These excellent references contain sound, useful information that should help the residents of such areas minimize the losses caused by extreme wind or rising water. Many of these publications are free. Some, government publications, are paid for by your taxes, so why not use them? The following references are recommended to those readers who wish to investigate further the subject of hurricane-resistant construction (see also references 72 and 79).

Building a home (general)

92. *Coastal Design: A Guide for Builders, Planners, and Homeowners*, by Orrin H. Pilkey, Jr., Orrin H. Pilkey, Sr., Walter D. Pilkey, and W. J. Neal, 1983. A detailed companion volume and construction guide expanding on the information outlined in this text. Chapters include discussions of shoreline types, individual residence construction, making older structures storm-worthy, high-rise buildings, mobile homes, coastal regulations, and the future of the coastal zone. To be published by Van Nostrand-Reinhold Company, New York, NY.

93. *Design and Construction Manual for Residential Buildings*

Appendix C. Useful references

in Coastal High Hazard Areas, prepared by Dames and Moore for HUD on behalf of the Federal Emergency Management Agency (FEMA), Federal Insurance Administration, 1981. A guide to the coastal environment with recommendations on site and structure design relative to the National Flood Insurance Program. The report includes design considerations; examples; construction costs; appendices on design tables, bracing, design worksheets, and wood preservatives; and a listing of useful references. The manual is available from the Superintendent of Documents, U.S. Government Printing Office, Washington, D.C. 20402 (publication number 722-967/545) or contact a FEMA office.

94. *Standard Building Code* (1979, previously known as the *Southern Standard Building Code* and still frequently referred to by that name). Available from Southern Building Code Congress International, Inc., 900 Mountclair Road, Birmingham, AL 35213.

95. *Dwelling House Construction Pamphlet of Southern Standard Building Code.* Conforms to the Southern Standard Building Code and applies only to dwellings of wood-stud or masonry-wall construction. Available from Southern Building Code Congress or Southern Building Code Publishing Company (address under previous reference).

96. *The Uniform Building Code.* Available from International Conference of Building Officials, 5360 South Workman Mill Road, Whittier, CA 90601.

97. *Building Construction Checklist for the Texas Coast and Shoreline.* An excellent brochure for persons buying or upgrading shoreline properties. Available from your local building officials or the Texas Coastal and Marine Council, P.O. Box 13407, Austin, TX 78711.

98. *Model Minimum Hurricane-Resistant Building Standards for the Texas Gulf Coast*, by the Texas Coastal and Marine Council, P.O. Box 13407, Austin, TX 78711.

99. "Hurricane Exposes Structure Flaws," by Herbert S. Saffir. In *Civil Engineering Magazine*, February 1971, pp. 54-55. Available from most university libraries.

100. *Potential Wind Damage Reduction Through the Use of Wind-Resistant Building Standards.* Should be of interest to builders of new structures. Available from the Texas Coastal and Marine Council, P.O. Box 13407, Austin, TX 78711.

101. *Estimating Increased Building Costs Resulting from Use of a Hurricane-Resistant Building Code.* Should be of interest to builders of new structures. Available from the Texas Coastal and Marine Council, P.O. Box 13407, Austin, TX 78711.

102. *Structural Failures: Modes, Causes, Responsibilities*, 1973. See especially the chapter entitled "Failure of Structures due to Extreme Winds," pp. 49-77. Available from the Research Council on Performance of Structures, American Society of Civil Engineers, 345 East 47th Street, New York, NY 10017.

103. *Wind-Resistant Design Concepts for Residences*, by Delbart B. Ward. Displays with vivid sketches and illustrations con-

struction problems and methods of tying structures down to the ground. Considerable text and excellent illustrations devoted to methods of strengthening residences. Offers recommendations for relatively inexpensive modification that will increase the safety of residences subject to severe winds. Chapter 8, "How to Calculate Wind Forces and Design Wind-Resistant Structures," should be of particular interest to the designer. Available as TR-83 from the Defense Civil Preparedness Agency, Department of Defense, The Pentagon, Washington, D.C. 20301; or the Defense Civil Preparedness Agency, 2800 Eastern Boulevard, Baltimore, MD 21220.

104. *Interim Guidelines for Building Occupant Protection from Tornadoes and Extreme Winds*, TR-83A, and *Tornado Protection—Selecting and Designing Safe Areas in a Building*, TR-83B. These are supplement publications to the above and are available from the same address.

105. *Hurricane-Resistant Construction for Homes*, by T. L. Walton, Jr., 1976. An excellent booklet produced for residents of Florida, but equally as useful to those of the Texas coast. A good summary of hurricanes, storm surge, damage assessment, and guidelines for hurricane-resistant construction. The booklet presents technical concepts on probability and its implications on home design in hazard areas. There is also a brief summary of federal and local guidelines. Available from Florida Sea Grant Publication, Florida Cooperative Extension Service, Marine Advisory Program, Coastal Engineering Laboratory, University of Florida, Gainesville, FL 32611.

106. *Guidelines for Beachfront Construction with Special Reference to the Coastal Construction Setback Line*, by C. A. Collier and others, 1977. Report No. 20, available from Florida Sea Grant Publication, Florida Cooperative Extension Service, Marine Advisory Program, Coastal Engineering Laboratory, University of Florida, Gainesville, FL 32611.

Wood structures

107. *Houses Can Resist Hurricanes*, by the U.S. Forest Service, 1965. An excellent paper with numerous details on construction in general. Pole-house construction is treated in particular detail (pp. 29-45). Available as *Research Paper FPL 33* from Forest Products Laboratory, Forest Service, U.S. Department of Agriculture, P.O. Box 5130, Madison, WI 53705.

108. *Wood Structures Survive Hurricane Camille's Winds*. Available as *Research Paper FPL 123*, October 1969, from Forest Products Laboratory (address under previous reference).

109. *Wood Structures Can Resist Hurricanes*, by Gerald E. Sherward, 1972. See *Civil Engineering Magazine*, September 1972, pp. 91-94. Available from most university libraries.

Masonry construction

110. *Standard Details for One-Story Concrete Block Residences*, by the Masonry Institute of America. Contains nine fold-out drawings that illustrate the details of constructing a concrete-

block house. It presents principles of reinforcement and good connections that are aimed at design for seismic zones, but these apply to design in hurricane zones as well. Written for both layman and designer. Available as Publication 701 from Masonry Institute of America, 2550 Beverly Boulevard, Los Angeles, CA 90057.

111. *Masonry Design Manual*, by the Masonry Institute of America. An 8.5-inch by 11-inch, 384-page manual that covers all types of masonry including brick, concrete block, glazed structural units, stone, and veneer. Very comprehensive and well presented. This book is probably of more interest to the designer than to the layman. Available as Publication 601 from the Masonry Institute of America (address under previous reference).

Pole-house construction

112. *Pole House Construction*. Available from the American Wood Preservers Institute, 1651 Old Meadows Road, McLean, VA 22101.

113. *Elevated Residential Structures, Reducing Flood Damage through Building Design: A Guide Manual*, prepared by the Federal Insurance Administration, 1976. An excellent publication outlining the threat of floods and the necessity for proper planning and construction. Illustrates construction techniques. Includes a glossary, worksheets for estimating building costs, and a list of additional references. Available from the U.S. Federal Insurance Administration, Department of Housing and Urban Development, 451 7th Street, S.W., Washington, D.C. 20410; or order publication 0-222-193 from the Superintendent of Documents, U.S. Government Printing Office, Washington, D.C. 20402.

Mobile homes

114. *Protecting Mobile Homes from High Winds*, TR-75, prepared by the Defense Civil Preparedness Agency, 1974. An excellent sixteen-page booklet that outlines methods of tying down mobile homes and means of protection such as positioning and wind-breaks. Publication 1974-0-537-785, available free from the Superintendent of Documents, U.S. Government Printing Office, Washington, D.C. 20402; or from the U.S. Army, AG Publications Center, Civil Defense Branch, 2800 Eastern Boulevard (Middle River), Baltimore, MD 21220.

115. *Suggested Technical Requirements for Mobile Home Tie Down Ordinances*, TR-73-1, prepared by the Civil Defense Preparedness Agency, July 1974. Should be used in conjunction with TR-75 noted above. Available from U.S. Army Publications Center, Civil Defense Branch, 2800 Eastern Boulevard (Middle River), Baltimore, MD 21220.

Federal legislation

The following are specific references to the federal legislation mentioned in chapter 5. The first reference is to where these can be found in the *United States Code*. The second refers to their place in *Statutes at Large*.

116. National Flood Insurance Act of 1968 (P.L. 90-448), enacted on August 1, 1968. (1) 42 U.S.C. sect. 4001 et seq. (1976). (2) 82 Stat. 476, Title 13.

117. Federal Water Pollution Control Act Amendments of 1972 (P.L. 92-500), enacted on October 18, 1972. (1) 33 U.S.C. sect. 1251 et seq. (1976). (2) 86 Stat. 816.

118. Marine Protection, Research and Sanctuaries Act of 1972 (P.L. 92-532), enacted on October 23, 1972. (1) 33 U.S.C. sect. 1401 et seq. (1976). (2) 86 Stat. 1052.

119. Flood Disaster Protection Act of 1973 (P.L. 92-234), enacted on December 31, 1973. (1) 42 U.S.C. sect. 4001 et seq. (1976). (2) 87 Stat. 975.

120. Water Resources Development Act of 1974 (P.L. 93-251), enacted on March 7, 1974. (1) 16 U.S.C. sects. 4601-13, 4601-14, 460ee (1976); 22 U.S.C. sect. 275a (1976); 33 U.S.C. sects. 59c-2, 59k, 579, 701b-11, 701g, 701n, 701r, 701r-1, 701s, 709a, 1252a, 1293a (1976); 42 U.S.C. sects. 1962d-5c, 1962d-15, 1962d-16, 1962d-17 (1976). (2) 88 Stat. 13, Title 1.

Appendix D. Field trip guides

Mustang and North Padre Islands

0.0 *Stop 1. Port Aransas ferry.* The rapidly growing town of Port Aransas can be reached by two routes: (1) driving the entire length of Mustang Island from Corpus Christi (25 miles) via the Kennedy Causeway or (2) crossing the Corpus Christi Ship Channel on the toll-free ferry from Harbor Island. These roads also serve as evacuation routes but neither is completely reliable at all times. The shortest route to the mainland depends on the ferries, which stop running when high waves and strong currents pose a threat to safe operation. The longer alternate route can also be cut off when Park Road 53 is flooded or washed out in the Corpus Christi Pass-Packery Channel area.

The sediments that make up Harbor Island were deposited as sand bars and mud flats associated with Aransas Pass, the adjacent tidal inlet (fig. 4.30). Before it was modified, the area between Mustang Island and Harbor Island was a shallow submerged sand flat. Horses and wagons could easily cross the sand flat because the water was only a few feet deep. At present, water depths approach 45 feet where the Corpus Christi Ship Channel was dredged, and surface elevations on nearby Harbor Island have been raised for the facilities where oil is stored and for the fabrication yard where offshore drilling and production platforms are made.

0.5 Stop light. From the ferry landing proceed east (straight ahead) toward the beach. The two-story wooden building on the right is the Tarpon Inn, one of the oldest buildings in Port Aransas. Tarpon was the name of the early settlement on the north end of Mustang Island. A number of buildings have been located on this site since the first inn was opened in 1886. After it burned in 1900, the inn was rebuilt in 1904. That building was also short-lived; it was destroyed by the 1919 hurricane. The present inn was built in the early 1920s. Notice how the ground floor is elevated to minimize flood damage. The strength of the building comes from the concrete pillars at each corner of the rooms. The pillars extend from beneath the ground surface to the roof and they provide support as well as lateral stability during strong winds (see chapter 6).

0.6 Stop street. Proceed east (straight ahead) on Cotter.

1.1 The high, densely vegetated dunes on the right marked the landward limits of the beach in the late 1800s. Now the dune ridge is more than 1,000 feet from the present-day shoreline. The seaward advance of the shoreline was partly caused by the southerly migration of Aransas Pass and partly by the subsequent construction of the jetties.

1.2 On the left is the Marine Science Institute, a research and teaching laboratory operated by The University of Texas at Austin. Visitors are invited to tour the facilities and to see the displays of marine life and research equipment. Notice also the types of construction. The oldest buildings are made of wood and raised on wood pilings above the expected flood heights. The newest buildings are also raised but on concrete pilings. In some buildings ground-level offices and labs are enclosed by "break-away walls." These walls are designed to give in if flood waters and floating debris exert excessive forces on the exterior walls.

1.5 *Stop 2. South jetty.* Park on the beach near the granite jetty. Where you are standing was once under water in the Gulf of Mexico and part of the sand shoals formed by Aransas Pass. But man completely altered the area by building the jetties and deepening the tidal inlet.

Between 1866 and 1899 Aransas Pass and the southern tip of San Jose Island migrated southward nearly 4,500 feet. As a result, the lighthouse which formerly marked the inlet entrance (fig. 4.30) was stranded behind San Jose Island. Several attempts were made to stabilize the channel but none were entirely successful until the north and south jetties were completed in 1916. The northern end of Mustang Island grew seaward as Aransas Pass eroded southward, and the beach continued to grow during and after jetty construction. By 1958 the beach had reached its present position where it has remained relatively stable.

Return to your car and drive south (away from the jetty), along the beach. Please observe the speed limit and traffic lanes and watch carefully for people crossing the beach. As you pass the bath houses, notice how they are elevated on concrete pilings that offer minimum resistance to storm waves.

2.1 *Stop 3. Horace Caldwell Pier.* This fishing pier has been damaged by many storms; most recently it was destroyed by Hurricane Allen (1980), which flooded the area with nearly 9 feet of water. Turn west (right) on the beach access road (Beach Street) and drive to Park Road 53. Turn south (left) and proceed 3.1 miles to Beach Access Road 1.

4.9 As you drive along the vegetated barrier flat, you can see the well-developed dune ridge on the east.

5.9 Beach Access Road 1. Turn east (left) and drive 0.1 mile to the beach.

6.0 Old-time residents of Mustang Island built on high ground immediately landward of the dunes or on the dune ridge itself. These wise coastal residents respected the coastal hazards and did not lower the dunes. Notice also that the integrity of the dunes has not been altered by building single-family houses on the dune ridge. Turn south (right) and drive along the beach. At present this barrier segment is sparsely developed (fig. 1.3); however, plans are under way to construct several resort communities. The drive offers a view of a less crowded beach and a natural beach-dune system.

12.0 **Stop 4. Lost Colony.** The name of this luxurious resort was nearly prophetic when gulfside residences of the Lost Colony were undermined by Hurricane Allen (1980). As a result of severe dune erosion during the storm, the support pilings were exposed and the structures were perched out over the backbeach. Sand was subsequently hauled in to replace the dunes, and then sand fences were erected to trap additional sand seaward of the buildings. Continue driving south along the beach.

14.0 Turn west (right) on Beach Access Road 2. Drive 0.3 mile and turn south (left) on Park Road 53.

14.3 Looking toward the bay you will see surface equipment that supports the nearby oil-field activities. Exploration and production of important energy resources beneath Mustang Island have been continuous since the mid-1950s with only minor or temporary disruptions in the island's ecosystem. Several fields were developed prior to the strict environmental regulations that exist today. Even within those fields marine grasses have recolonized the disturbed areas, and the location of pipelines laid through the dunes is virtually impossible to detect without some visual assistance such as a pipeline marker or a warning sign.

15.3 The vegetated sand hills on both sides of the road are stable dunes that were active during the early 1930s. At that time much of this area was an active dune field which had been caused by the severe droughts. The lack of rainfall and vegetation allowed the dunes to migrate from the beach across the island. Subsequent wet years have supported the growth of dense vegetation that has stabilized the dunes.

16.4 **Stop 5. Corpus Christi Fish Pass.** Pull off the road and park near the channel. The fish pass is an artificial channel that was dredged across Mustang Island in 1972. The channel was opened to allow the free migration of marine fauna (primarily fish and shrimp) between the Gulf and Corpus Christi Bay. The same purpose was served by a tidal inlet located in the southern part of Corpus Christi Bay until it closed in 1929. The fish pass, which also has a tendency to silt and close, is a favorite fishing spot for visitors to Mustang Island State Park. Return to your car and proceed south on Park Road 53.

18.8 **Stop 6. Corpus Christi Pass.** Carefully pull off on the side of the road and park your car. This broad sand flat, which is partly submerged and partly exposed, marks the northern limit of a tidal inlet that formerly opened into Corpus Christi Bay. The inlet had a history of migration, and old maps show the southward movement of the inlet in the mid to late 1800s. Newport Pass is also a remnant of the tidal inlet, as is Packery Channel, the southernmost position of the inlet shown on historical documents. Inlet migration was continuous throughout this broad area and dunes are still sparse. At present, the passes are periodically reopened and act as washover channels; the former sand shoals are maintained as washover fans. Flood waters move rapidly through these washovers during hurricanes and erode whatever is in their path including the road and the bridge approaches. The differences in road pavement along this highway segment are signs that the road has been replaced after it was washed out. Park Road 53 was damaged so severely by Hurricane Allen that major repairs were required before it could be used. Return to your car and continue driving south on Park Road 53.

20.2 Newport Pass.

21.4 High bridge over Packery Channel. The channel got its name from the meat packeries that were located on the banks of the tidal inlet. Cattle that grazed on the barrier islands in the late 1800s were processed at this site and the carcasses were dumped into the active tidal inlet.

21.9 Stop light. Turn north (right) on Park Road 22. Go 0.9 mile and turn left on Aquarius Street. Padre Isles development is a planned resort community for permanent residences of medium to high density. Residents of this community must cross the John F. Kennedy Causeway to reach the mainland. The causeway bridge over the intracoastal waterway is well above flood levels but the road across Laguna Madre is only a few feet above sea level and is periodically inundated.

23.1 Residential intersection. Turn east (left) on Jackfish Street and go 0.2 mile to the boat ramp and park your car.

23.3 **Stop 7. Bulkheaded finger canals.** Surface elevations of the building sites were increased by material dredged from the adjacent finger canals. The canals are broad, deep, and interconnected so that tidal currents and wind-blown currents allow the marina water to circulate with water from Laguna Madre. Furthermore, the community has its own sewage system. Hence the problem of stagnant and eutrophic finger canals with poor water quality is partly avoided. The bulkheads act as retaining walls for the fill material above sea level and protect the banks from erosion by boat wakes and storm waves. Return to your car and go back to Park Road 22. Turn south (right) and go 2.5 miles to the entrance to Padre Balli Park (Nueces County Park). Drive slowly along the park road to the beach.

26.7 **Stop 8. Bob Hall Pier.** Park your car and look around you. To the north, the beach is lined by hotels and condominiums but to the south, the barrier is undeveloped. Bob Hall Fishing Pier has survived some storms and has been damaged by others; it was completely destroyed by Hurricane Allen (1980), which generated a storm surge of about 9 feet in this area. The seawalls adjoining the nearby condominiums also were damaged (fig. 4.35) even though the storm crossed the coast more than 80 miles to the south. Return to your car and go back to Park Road 22. Turn south (left) and proceed toward Padre Island National Seashore.

28.5 Some dunes are still actively migrating across Padre Island (fig. 4.1) under the influence of strong onshore breezes.

35.2 Entrance to Padre Island National Seashore.

36.6 **Stop 9. Grassland nature trail.** Park your car in the designated area and take the self-guided tour. A clearly marked path with illustrations and explanations will assist you in learning more about the vegetated barrier flat, the most extensive environment of the Texas barriers. Return to your car and proceed south on Park Road 22.

37.3 Road to Bird Island Boat Basin. Turn west (right) and drive toward Laguna Madre. This road transects most of the barrier environments including the vegetated barrier flat, active back-island dunes, and intervening deflation troughs. The deflation troughs are identified by marsh vegetation and standing water.

37.8 Off in the distance (to the south) one can see the remnants of live-oak trees, evidence that woody vegetation flourished on the island under different climatic conditions.

38.2 The low active dunes to the south (left) of the road have been migrating toward the lagoon at nearly 75 feet per year. The sand blown from the back island into Laguna Madre has helped keep the lagoon a shallow body of water with a sandy floor. Marine grasses and other algae thrive in the clear, warm water of the lagoon. These grasses die during the winter and are washed onto the beach where they form thick mats interlayered with sand and shell debris.

38.8 **Stop 10. Laguna Madre.** Park your car along the lagoon margin. Notice that the waves are extremely small and the sandy beach is relatively low and narrow. The shape of this beach and the energy of the waves are typical of lagoons and bays protected by the barriers. Return to your car and head back to Park Road 22. As you drive away from the lagoon notice that the ground elevations increase toward the dunes. This is one of the reasons why the safest building sites on a barrier island are immediately landward of the dune ridge.

Across from the intersection you can see the decaying remnants of a wooden corral and outbuildings that were used when Padre Island was a working cattle ranch. Ranching operations stopped shortly after the National Park Service acquired the property in 1962. Turn south (right) on Park Road 22 and drive to Malaquite Beach.

43.2 **Stop 11. Malaquite Beach.** Park your car in the visitors' parking area and tour the facilities. When you walk to the beach you will see that the complex that houses the bathhouse and visitors' center is located seaward of the dunes. As pointed out in chapter 2, building in the backbeach is a hazardous proposition and Malaquite Beach is no exception. Even though the visitors' center is raised far above the flood levels experienced during Hurricane Allen (11.5 feet), the buildings suffered damage that cost more than $500 thousand to repair. Consequently, the revised master plan calls for the eventual demolition of the present visitors' center, construction of new facilities landward of the dune ridge, and reestablishment of the dunes. The firsthand experience with coastal processes and a lower visitor rate than expected have prompted a reevaluation of the park's development. The revised master plan definitely favors construction and park utilization that are compatible with the natural setting and coastal processes.

South Padre Island

0.0 **Stop 1. Port Isabel Lighthouse.** Park your car and walk to the lighthouse. The climb to the top (all seventy-four steps) is well worth the panoramic view of the mainland (Rio Grande delta), Laguna Madre, and South Padre Island. Since it was built in 1851, the lighthouse has served as a landmark, observation post, and a navigational aid to boat traffic through Brazos Santiago Pass. Unlike some lighthouses built on rapidly eroding shores (such as the Cape Hatteras Lighthouse on the North Carolina coast), the Port Isabel Lighthouse occupies a relatively safe site that is protected from storm waves. Nevertheless, its usefulness is threatened but not by coastal processes. The lighthouse stands more than 70 feet above ground level, and for more than a century it was the highest structure in this part of Texas. Now the tower and light room are dwarfed by high-rise condominiums and hotels that signal the beginning of a new chapter in the history of Port Isabel and South Padre Island.

0.1 Drive east across the Queen Isabella Causeway (Park Road 100) to South Padre Island, which is the most densely developed barrier island (fig. 1.2) on the Texas coast. The causeway is the route for evacuating the residents when the island is threatened by a storm. Even though the bridge is much higher than expected flood elevations, the approaches to the bridge are low and subject to flooding (fig. 4.5).

2.8 Intersection. Turn south (right) on Padre Boulevard.

3.2 **Stop 2. Old Queen Isabella Causeway.** Turn west (right) and proceed to the old causeway. This low, narrow structure was the transportation link to the mainland before the new causeway was built in the early 1970s. The building boom on South Padre and the concomitant increase in residents and visitors necessitated improving the road system to the island. The old causeway no longer carries traffic but it does support a pipeline that supplies fresh water to the island; it also serves as the nation's longest fishing pier. Return to Padre Boulevard and turn south (right).

4.1 **Stop 3. Coast Guard Station.** Turn east (right) and proceed to the new Coast Guard Station. Tours for visitors are only conducted on Saturdays and Sundays, so at other times please be as unobtrusive as possible. Notice that the lowest level of the headquarters building is designed so that flooding will not disrupt the base of operations. Although these facilities were designed to withstand the unusual forces exerted by storms, they suffered some damage in 1980 during Hurricane

Allen. Maximum flood elevations along the Gulf beach and in Laguna Madre were about 8 feet above sea level. Return to Padre Boulevard, turn south (right), and proceed to Isla Blanca Park. Bear left and enter the parking lot.

5.3 **Stop 4. Brazos Santiago Pass**. Park your car and walk to the beach next to the granite jetty. You are standing where a natural tidal inlet and sandy shoals existed until 1935. At that time dredging operations were conducted to widen and deepen the inlet, and the jetties were constructed to stabilize the channel. After construction of the jetties, the Gulf shoreline accreted rapidly and advanced seaward more than 1,000 feet. Since 1960 the broad sand beach adjacent to the north jetty has been relatively stable whereas the beach along the remainder of the island has continued to erode. Look northward along the shoreline (away from the jetty) and notice the curvature of the beach. The shoreline in front of the hotels and condominiums is landward of (indicating erosion) but parallel to the shoreline that existed before the jetties were constructed. The shoreline curvature has resulted from accretion near the north jetty and erosion along much of the beach that is now the site of expensive resort developments.

Walk northward along the beach and note the general lack of dunes except between the park pavilions and the hotels. Here one can see remnants of the foredune ridge that stood nearly 20 feet above the sea and provided storm protection before the dunes were leveled for building sites.

After examining the dunes you might want to visit the Pan American University Marine Biology Laboratory where aquatic plants and animals are on display. Return to your car and proceed northward, retracing your route back toward the causeway.

6.4 Turn east (right) and drive past the resort facilities and notice how the lowest levels of the buildings are used. The ground elevation around these structures has been raised mainly by lowering the dunes. Some building sites have been elevated by sand excavated from the island core. Even with the increased elevation the ground surface is still below expected flood levels for an extreme hurricane. All the buildings are on concrete pilings; some are raised above the land so flood waters will flow beneath the first floor, some use the ground level for parking, and some use that area for lobbies or permanent residences.

7.2 Turn north (right) on Padre Boulevard and drive 0.6 mile to Sheepshead. Turn east (right) and drive 0.1 mile to Gulf Boulevard. Turn left and proceed north on Gulf Boulevard which runs parallel to the beach.

8.5 To the right is a house built on the beach. The "single-lot" seawall that once protected the house from Gulf waves has failed, and the house is slowly deteriorating as the shoreline continues to erode. As you drive along Gulf Boulevard notice the different types of materials that have been used in constructing the buildings; also note the orientation of the buildings and their distance from the shoreline. The largest developments have one thing in common: they each have a seawall (fig. 4.40) designed to resist storm waves and resulting beach scour. Not all the building designs, however, take into account the strong lateral forces produced by high wind velocities (fig. 4.40).

The foundations and pilings of some of the buildings were exposed by Hurricane Allen in 1980, and nearly all the seawalls had to be repaired or replaced because they were severely damaged or destroyed during the storm. The South Padre beach is lined with multi-family dwellings not only because beach property is extremely valuable but also because single-family dwellings are safest when located landward of the dunes.

10.0 *Stop 5. Failed seawall*. Park your car and walk to the beach. Here you will see another failed seawall that was erected to protect private property. When it was constructed in 1962, the wall extended 1,500 feet along the beach; it was 8 feet high and 7.5 inches thick. This massive reinforced concrete structure was essentially destroyed by Hurricane Beulah in 1967. A previous seawall on the same property was destroyed by Hurricane Carla in 1961. The location of this seawall seaward of the beach is clear evidence that the shoreline is eroding and has been for many years.

10.2 Turn west (left) on Sunset. Drive 0.2 mile and turn north (right) on Padre Boulevard.

10.7 *Stop 6. Buildings in a washover channel*. Turn east (right) on White Sands and park your car. Because dunes are sparse or absent in washover channels, this building site was elevated using material dredged from the back-island sand flats (a washover fan); the surrounding area, however, is still extremely low and it is susceptible to flooding and washover by storm waves. A major channel was cut across the island at this location during Hurricane Beulah (fig. 4.3).

Nearly all the high-rise buildings on South Padre Island have been built since the early 1970s. Hence, they have not experienced a major storm. Some local residents claim that Hurricane Allen tested the durability of the developments. Such claims ignore two important facts: (1) the storm stalled and was greatly weakened before it moved onshore, and (2) the developed areas were located south of the storm center, not in the area of maximum wind speeds and surge heights. Return to your car and proceed north on Padre Boulevard.

11.7 From the road you can see the Gulf of Mexico (to the east) and Laguna Madre (to the west). The uninterrupted view tells you that (1) the island is narrow and that (2) you are driving across a washover area. The parallel arrangement of sand fences in the washover is designed to trap sand and form artificial dunes. Artificially nourished sand dunes may be desirable but they are also temporary. Larger dunes than these have been washed away during previous storms. In fact, this section of the road was washed out during Hurricane Allen, and it is likely that the sand fences and dunes will also be washed out during the next storm.

12.1 Large active sand dunes up to 20 feet high occur between the washover channels. This alternating pattern of washover channels and active dunes is repeated for the next 6 miles along the paved road. At times the barren sand dunes migrate onto the road and have to be removed by front-end loaders and dump trucks.

12.8 As you cross the next washover area notice that the color and texture of the pavement change. The new pavement is located where the road was washed out during Hurricane Allen and subsequently replaced. The land along this hazardous barrier segment has been subdivided and plans are being made to extend the development northward. Find a place to turn around and return southward to the park entrance and beach access road.

13.6 *Stop 7. American Veterans Park*. Turn east (left) and drive to the beach where you will find ample space for parking. This stop offers a view of the natural beach and dune system of a barrier island that is migrating landward. As discussed in chapter 2, this is accomplished by erosion along the Gulf shoreline and deposition (washover fans) along the lagoon margin. Most of the dunes are low, discontinuous, and sparsely vegetated; others are large, nearly barren masses of sand. All of these dunes are relatively young and have formed in a few years since the last major storm.

The shells and rock fragments on the beach also have a story to tell. The shells are a mixture of species that live in the Gulf and the lagoon. Many of the shells and rock fragments are rounded, abraded, and pitted, indicating that they are relatively old. The presence of many oyster shells on the beach indicates that the barrier is slowly moving landward over sediments that were deposited in protected bays or lagoons. Now these same sediments are exposed on the Gulf floor and are being eroded by the waves and tossed up on the beach.

Comparing old maps and aerial photographs shows that between 1867 and 1974 this beach eroded 1,400 feet, or at an average rate of 13 feet per year. The significance of this distance and erosion rate is underscored when one considers that the barrier at this location is only about 2,500 feet wide. Return to your car and drive south on Padre Boulevard.

14.8 Turn to the west (right) and park your car. The canals you see were dredged to provide homesites with adjacent boat facilities and access to Laguna Madre. The elevations of the homesites were raised by the material removed from the channels. The fill material and underlying barrier are composed of loose sand and shell fragments; these are easily eroded by waves (or boat wakes) or by runoff of floodwaters that inundate the area during storms. To prevent such erosion, the channels and lagoon shores are lined with bulkheads.

The canals are wide, relatively short, interconnected, and open to Laguna Madre. Because of these features, the canals are flushed by tidal and wind-driven currents. Pollution in these finger canals and degradation of water quality are minimized by the periodic exchange of canal and lagoon water as well as by the use of a centralized sewage system rather than septic tanks.

Index

access roads, 19–20, 158
accretion, shoreline. *See* barrier islands, widening
aerial photography, 152
anchoring, house. *See* construction
Aquarius Street, 178
Aransas Pass, 4, 7, 45, 94, 96, 97, 99, 101, 104, 119, 166, 167, 175, 176
    lighthouse, 96

back barrier flats, 18, 44
backmarshes. *See* marshes
barometric pressure, 25, 26, 124, 125–126, 139
barrier coast, 1, 3
barrier islands, 1, 12–14, 16, 17, 31, 35, 119, 163, 164
    dynamics, 20, 36, 121
    environments, 18–20
    evolution, 16–18
    migration, 13–15, 17, 23, 36, 56, 69, 76, 87, 152, 154–155
    origin, 12
    stability indicators, 40
    storm response, 20
    widening, 14, 18, 23, 49, 181
Bay City, 162
Bay Harbor, 68
bays, 3, 5, 6, 18, 40, 43, 44, 109, 119, 179, 183
beach access. *See* access roads
Beach Access Road 1, 176

Beach Access Road 2, 177
beaches, 163, 164
    artificial, 25, 40
    dynamic equilibrium, 20, 36
    erosion. *See* barrier islands, migration
    replenishment, 25, 29–30, 37, 44, 63, 101, 154, 181
    shape, 19–21, 22, 23, 30, 31, 36
    stabilization. *See* shoreline engineering
    storm response, 20, 29
Beaumont, 4, 162
Big Shell Beach, 17, 24
Bird Island Cove, 66
Bob Hall Fishing Pier, 178
Boca Chica
    Bay, 112
    Beach, 112
    Island, 112
    Pass, 109
Bolivar Peninsula, 4, 7, 12, 15, 42, 44, 45, 46, 49, 56–57, 58–59, 60–61, 62
Bolivar Roads, 4, 7, 56, 57, 60, 166
Brazoria County, 4, 162
Brazos Island, 4, 5, 119, 167
Brazos River, 4, 5, 7, 16, 71, 75, 76
    delta, 71, 76, 166
Brazos Santiago Pass, 4, 105, 109, 113, 114, 119, 180, 181
bridges, 152

Brown Cedar Cut, 4, 5, 77, 78, 81, 166
Brownsville, 4, 10, 29, 162
Bryan Beach, 74–75
    State Park, 71, 75
building codes, 117, 121, 123, 129, 131, 147, 153, 170–172
buildings
    apartment, 122, 137–138, 142
    high-rise, 37, 43, 100, 101, 105, 114, 142–145, 170, 178, 181
    mobile homes, 101, 122, 139–142, 158, 170, 173
    modular, 145, 147
    single-family, 122, 139, 176, 182
bulkheads, 31, 34, 44, 86, 109, 178
Bureau of Economic Geology, 2, 25

Calhoun County, 4
Cameron County, 4, 105, 106, 108
canals. *See* finger canals
Caplen, 58
Cape May, N.J., 34–35
Carancahua Cove, 67
causeways, 152
    wooden, 7, 180
Cedar Bayou, 4, 90, 92, 93
Cedar Lakes, 74, 76, 77, 78–79
Chambers County, 4
channels. *See* overwash

civil preparedness, 153
Coastal Barrier Resources Act, 116–117
coastal development, 1, 3, 8, 39, 62, 152–159, 169
    future, 24, 109, 115–120
    history, 6–8
    safe, unsafe. *See* construction, site safety
coastal hazards, 27, 124, 127, 166–168
Coastal Public Lands Management Act, 119–120, 153
coastal surveys, 6
Coastal Zone Management Act, 169, 170
Cold Pass, 68, 71
Colorado River, 4, 5, 16, 26, 77, 80, 82, 86
condominiums. *See* buildings, high-rise
Coney Island, N.Y., 27
conservation, shoreline, 36
construction, 45, 122, 131, 170–173
    anchoring, 126, 128, 129, 132, 133, 134, 137–138, 144, 181
    brick, 136
    concrete block, 136, 137, 145
    design, 123, 126
    masonry, 136, 172–173
    pole or stilt, 126, 128, 129, 131, 132, 173
    roof, 129, 132, 134, 138
    shape, 123, 136, 137
    site safety, 122, 131–132
    slabs, 132, 144
    strengthening, 125, 129, 135, 137
    walls, 131, 134, 136, 139, 176
    wood, 126, 172

continental shelf, 13, 16, 30
Corps of Engineers. *See* U.S. Army Corps of Engineers
Corpus Christi, 4, 10, 11, 29, 30, 100, 118, 143, 162, 163, 175
    Bay, 12, 98, 104, 177
    Pass, 101, 102–103, 104, 177
county commissioners' courts, 118–119
Crystal Beach, 41, 56–57, 61
currents. *See* longshore currents
cyclones. *See* storms, tropical

dams, 16, 17, 23, 109
Dana Cove, 67
Dead Caney Lake, 9
Delehide Cove, 67
Del Mar Beach, 112
disaster assistance, 153–154
dredging. *See* beaches, replenishment
dune buggies, 19, 23, 44
dunes, 5, 17, 18, 21, 27, 35, 36, 40, 43, 50, 53, 76, 80, 87, 92, 97, 100, 105, 109, 154, 175, 177, 178, 179, 181, 183
    artificial, 19, 182
    primary, 41, 44
    protection, 118–119
    stability, 19, 40, 41, 42, 176
dynamic equilibrium. *See* beaches

East Bay, 56, 57
East Beach, 63, 65

Eckert Bayou, 67
elevations, 18, 28, 40–41, 49, 50, 69–70, 77, 105, 109, 132
Eleven-Mile Road, 67, 69, 70
erosion, 1, 18
    due to engineering, 17, 20, 24, 25, 29, 34
    due to sea-level rise, 17–18, 23, 39, 53, 56, 57, 70, 71, 77, 80, 86, 87, 92, 97, 101, 105, 109, 114, 176
    indicators, 24
    rates, 24, 40
escape routes. *See* evacuation
evacuation, 19–20, 28, 48–49, 50, 53, 57, 62, 63, 71, 77, 80, 97, 100, 105, 109, 114, 132, 139, 150
excavation of sand. *See* sand, excavation
exploration
    French, 6
    Spanish, 6, 160

Farmers Home Administration mortgage loans. *See* mortgage loans
Federal Emergency Management Agency, 115, 121, 171
Federal Flood Hazard Map, 119
Federal Housing Administration mortgage loans. *See* mortgage loans
federal legislation, 115–120, 173–174
Federal Water Pollution Control Act, 117
    amendments, 158

ferries, 62
    Bolivar Roads, 57, 62
    Harbor Island, 100, 175
    Port Aransas, 49, 175
field trips, 3, 175–184
finger canals, 44, 46–48, 56, 57, 70, 101, 104, 156, 178, 183–184
fires, 143, 145
flood insurance. See National Flood Insurance Program
Flood Insurance Rate Map, 115
Follets Island, 4, 68, 71, 72–73
Fort Esperanzas, 7
Fort Louis, 6
fossils. See seashells
foundation, house. See construction, anchoring
Freeport, 10, 26, 29, 63, 71, 72, 76, 162
    Harbor, 7

Galveston, 6, 10, 11, 26, 34, 39, 49, 62, 118, 160, 161, 162, 163
    Bay, 12
    County, 4
    Island, 4, 7, 8, 13, 15, 24, 44, 46, 64–65, 66–67, 68, 70, 166, 168
    jetties, 57, 63
    seawall, 1, 15, 24, 29, 31, 33, 36, 40, 45, 63, 64, 66, 69, 166
Gangs Bayou, 67
geologic information 154, 162–163
Gilchrist, 41, 56, 59

glaciers, 12
Godfrey, Dr. Paul, 16, 19
Goldsmith, Dr. Victor, 30
Graber, Peter C., 115
Greens Bayou, 80, 85, 86
groins, 21, 23, 30–32, 34, 50
groundwater. See water, ground
Gulf Boulevard, 181
Gulf of Mexico, 3, 5, 9, 25, 26, 56, 86, 96, 105, 160, 176, 177, 182

Harbor Island, 96, 100, 104, 175
    lighthouse, 9, 97
harbors. See ports
Harlingen, 162
High Island, 4, 5, 53, 54, 55, 56, 57, 62
Horace Caldwell Pier, 176
Horseshoe Lake, 60
house selection guidelines, 126, 131
houses. See construction
Houston, 4, 29, 63, 163
Hurricane
    Abby, 10
    Allen, 39, 101, 104, 105, 109, 176, 177, 178, 179, 180–181, 182, 183
    Audrey, 57, 62
    Beulah, 10, 11, 43, 101, 105, 109, 161, 182
    Camille, 21, 39, 45, 46, 125–126, 136, 139
    Candy, 10
    Carla, 10, 11, 16, 26, 27, 42, 53, 56, 57, 62, 63, 69, 70, 71, 76, 77, 80, 86, 92, 100, 161, 182

    Carmen, 10, 48
    Celia, 10, 11, 101, 161, 163–164
    Cindy, 10, 62
    Debra, 62
    Edith, 10
    Eloise, 114, 142–143, 144
    Felice, 10, 62
    Fern, 10
    Hazel, 161
    of 1900, 1, 39, 63, 69, 161
hurricanes, 3, 48, 63, 126, 134, 148–151, 155–156, 159, 160–162, 172, 181
    defined, 25–26
    destruction, 6, 26–28, 122, 129, 171
    forces, 26, 123, 125
    frequency, 11
    history, 9–11, 160–162
    origin, 25–26
    precautions, 149

Indianola, 6
inlets. See passes
insurance. See National Flood Insurance Program
Interstate Highway, 45, 63
Intracoastal Waterway, 9, 56, 57, 58–59, 60–61, 72–73, 78–79
Isla Blanca Park, 109, 113, 181
islands. See barrier islands

Jackfish Street, 178

Jefferson County, 4
jetties, 1, 7, 18, 23, 30–31, 45, 56, 57, 60, 63, 76, 80, 86, 97, 99, 109, 176, 181
John F. Kennedy Causeway, 49, 100, 175, 178
Jumbile Cove, 66

Karankawas tribe, 6
Kenedy County, 4
Kingsville, 162, 163
Kleberg County, 4

lagoons. See bays
Laguna Madre, 7, 91, 97, 101, 102, 104, 105, 110–111, 113, 178, 179, 180, 181, 182, 183
land acquisition. See property rights
land-use regulations, 3, 50, 105, 115–120, 157, 169–170
La Punta Larga, 10
LaSalle, Robert, 6
Lavaca Bay, 6, 26
littoral drift. See longshore currents
longshore currents, 12, 21, 31
Lost Colony, 177
Lubbock, 143
Lydia Ann Channel, 96

Maggies Cove, 66
Malaquite Beach, 179–180
Mansfield Channel, 4, 45, 105, 107, 119, 167
maps, 152, 156–157, 183
Marine Protection Research and Sanctuaries Act, 117

marshes, 5, 15, 16, 44–45, 50, 53, 87, 92
 buried, 44
 filled, 46
 Spartina, 19
Matagorda
 Air Force Base, 7, 86, 87, 88
 Bay, 6, 78, 86, 167
 Club, 84
 County, 4
 Island, 4, 7, 13, 19, 87, 88–89, 90–91, 92, 97, 118
 Peninsula, 4, 15, 16, 18, 19, 24, 27, 45, 86, 166
  East, 77, 80, 81, 83
  West, 80, 82, 84–85
 Ship Channel, 4, 45, 84, 86
Miami Beach, 34
migration. See barrier islands
mobile homes. See buildings
Monmouth Beach, N.J., 34, 38
mortgage loans, 115
Mud Island, 94–95
Mustang Island, 4, 7, 8, 15, 18, 24, 45, 46, 49, 96, 97, 98–99, 100, 101, 102–103, 104, 117, 163, 167, 175–179
 State Park, 97, 98, 177

National Academy of Science, 23
National Flood Insurance Program, 50, 115, 116, 126, 156
National Oceanic and Atmospheric Administration, 49
National Park Service, 19, 179

New Jerseyization, 1, 8, 9, 16, 17, 29, 31, 33, 37, 38, 39
Newport Pass, 101, 103, 177, 178
North Padre Island, 7, 18, 29, 40, 41, 48, 49, 101, 102–103, 104–105, 163, 167, 175–179
North Pass, 92, 94, 97
Nueces County, 4, 119, 178

Offatt Bayou, 64
one-hundred-year floods, 115, 116, 125
Open Beaches Act, 118
orthophoto maps, 152
overwash, 16, 18, 19, 27, 40, 42, 43, 50, 53, 57, 58, 60, 67, 70, 76, 79, 80, 81, 83, 84, 86, 88, 90, 92, 105, 108, 109, 152, 178, 181, 182, 183
 fans, 18, 28, 182

Packery Channel, 4, 48, 101, 103, 104, 175, 177, 178
Padre Balli Park (Nueces County Park), 178
Padre Boulevard, 180, 181, 182, 183
Padre Island, 5, 6, 16, 17, 24, 26, 43, 46, 104, 105, 167, 179
 National Seashore, 4, 17, 102–103, 160, 164, 178, 179
Panama City, Fla., 114, 143
Pan American University Marine Biology Laboratory, 181
Panther Point, 91, 92
Park Road 53, 175, 176, 177, 178
Park Road 100, 180
Park Road 22, 178, 179

parks, 157, 164–165. *See also* field trips
Pass Cavallo, 4, 84, 86, 87, 89, 92, 166
passes, 2, 5, 31, 50, 56, 71, 77, 87, 92, 97, 101, 178
   artificial, 45
   former, 48
   migration, 45, 177
   potential, 4
peninsulas, definition, 3
piers. *See* construction, anchoring
piles. *See* construction, anchoring
poles. *See* construction, pole or stilt
pollution, 46, 48
Port Aransas, 4, 6, 7, 26, 49, 99, 100, 101, 175
Port Arthur, 162
Port Bolivar, 6, 62
Port Isabel lighthouse, 180
Port Lavaca, 162, 163
Port Mansfield, 105
ports, 1, 6, 7, 29, 31, 45, 100
power failures, 28, 143
Power Lake, 87, 91, 92
Pringle Lake, 88
property rights, 156, 158

Queen Isabella Causeway, 49, 109, 113, 180
Quintana, 72
   Beach, 71

recreation, 157, 164–165. *See also* field trips
Refugio County, 4
remote-sensing imagery, 152

Rio Grande, 4, 5, 16, 26, 39, 112, 114, 119, 161
   delta, 105, 109, 167, 180
Rollover Bay, 56, 59
Rollover Pass, 45, 53, 56, 57, 59

Sabine Pass, 4, 5, 7, 26, 39, 50, 51, 53, 57, 119, 166
Sabine River, 4
St. Josephs Island. *See* San Jose Island
salt marshes. *See* marshes
San Bernard National Wildlife Refuge, 74
San Bernard River, 16, 71, 74, 76
San Jose Island, 4, 13, 87, 92, 93, 94–95, 96, 97, 118, 166, 176
San Luis Pass, 4, 68, 70, 71, 166
sand, 18, 50, 105, 125, 144
   excavation, 70, 118
   fencing, 44, 100, 177, 182
   removal, 105, 109, 118
   supply, 16, 17, 20–21, 25, 27, 29–30, 31, 36, 38, 41, 44, 69, 86
sand bars. *See* spits
sanitation, 155, 158
Sargent Beach, 9, 29, 30, 46, 77, 78–79, 80, 167
Sea Island, 66
Sea Rim State Park, 50, 52, 53, 54–55
sea-level changes, 12, 13–14, 18, 39
seashells, 15, 17, 23, 41–42, 53, 77, 183
seawalls, 1, 21, 24, 29, 30–31, 33–35, 50, 101, 104, 109, 178, 181, 182
septic tank systems, 44, 46, 50, 57, 158, 168–169, 184

sewage, 46, 50, 101, 117, 156, 158, 184
Shamrock Island, 98
Shore and Beach (magazine), 115
shoreline engineering, 3, 29, 35, 36, 37–38
shoreline retreat. *See* erosion
shoreline stabilization, 24, 29, 36, 37, 45, 165–166
site analysis, 40, 50, 121–122, 131, 169
Small Business Administration mortgage loans. *See* mortgage loans
Snake Island Cove, 66
soils, 28, 40, 41, 50, 126, 158–159
South Bay, 112
South Harris County, 162
South Padre Island, 1, 4, 7, 8, 15, 18, 24, 29, 37, 40, 41, 43, 45, 48, 49, 106–107, 108, 109, 110–111, 112–113, 114, 117, 119, 167, 180–184
   Coast Guard Station, 180
Southern Standard Building Code, 118, 138, 171
Spartina patens, 16
spits, 6, 12, 21, 29, 86
Spring Bayou, 77, 80, 83
Standard Building Code. *See* Southern Standard Building Code
storm surge, 26, 27, 28, 49, 80, 101, 125, 172
storms, 7, 10, 21, 24, 26, 31, 36, 37, 38, 40, 43, 45, 62, 92, 97, 100, 101, 104, 105, 109, 121, 160
   frequency, 11
   survival, 162
   tropical, 25, 71, 76, 123, 136, 137

Sunset (road), 182
Surfside, 4, 5, 24, 30, 46, 71, 72, 76
surf zones, 12, 62
Sweetwater Lake, 67
swimming pools, 144
Sydnor Bayou, 64

Tarpon Inn, 175
terrain, 40, 42
Texas Coastal Program, 120, 153
Texas Department of
  Health, 117
  Highways and Public Transportation, 158
  Parks and Wildlife, 87
  Water Resources, 117
Texas Energy and Natural Resources Advisory Council, 120
Texas General Land Office, 119, 120
Texas State Highway
  No. 4, 109, 114
  No. 87, 53, 62
Tropical Storm Delia, 10, 57, 62

Uniform Building Code, 118, 171
U.S. Army Corps of Engineers, 24, 30, 56, 69, 86, 117, 159, 160
U.S. Department of Interior, 117
U.S. Geological Survey, 41

V-zones. *See* velocity zones
vegetation, 24, 41, 118, 159, 168, 179, 183
  patterns, 16, 17

stabilizing, 40, 44, 50, 100, 119, 177
velocity zones, 115–116
Veterans Administration mortgage loans. *See* mortgage loans
Veterans Park, American, 105, 110

washover channels. *See* overwash
waste disposal, 117. *See also* septic tank systems *and* sewage
water, 46, 50
  ground, 46
  problems, 168–169, 184
  resources, 159, 163
Water Resources Development Act, 117
waves, 15, 19, 20, 23, 26, 33, 39, 40, 42, 44, 57, 76, 104, 122, 124, 144, 164, 178, 180, 181
  energy, 12, 21, 36, 125
Whale Beach, N.J., 16
White Sands (road), 182
Willacy County, 4, 105, 106, 108
winds, 17, 123. *See also* hurricanes *and* storms

Yarborough Pass, 167